Coupe st: a b

CONTO

LIGNE DE CAPOT - PARTIE SUPÉRIEURE

910

642

226

161.5

150

180

24

537

320

216

80 64 42

6.00 18 - 4.700

AXE DU MOTEUR - PENTE 2%

32

R-450

34

30

R.500

452

b

½ Vue d'avant

POINT LE PLUS HAUT
DU CHASSIS PASSANT
L'AXE DES ROUES A

255

76

213

133

530

ENCOMBREMENT = 1m.6

SOL

VOIE DES ROUES AV AU SOL = 1m.349

722.5

238

R-145

1m570

RADIATEUR D'HUILE

664

VOIE DES ROUES AV AU CENTRE DES ROUES = 1m385

½ Coupe st: C d

POINT LE PLUS HAUT DU CHASSIS
PASSANT PAR L'AXE DES ROUES AR

263

80 DÉBATTEMENT

PENTE 14.61%

125

RÉGLAGE DES AMORTISSEURS AV

1m480

VOIE DES ROUES AR = 1m362

ENCOMBREMENT = 1m652

SOL

145 AUTOMOBILES DELAHAYE

Plan pour carrossier

Faster

BOOKS BY NEAL BASCOMB

Higher

The Perfect Mile

Red Mutiny

Hunting Eichmann

The New Cool

The Winter Fortress

The Escape Artists

Faster

Faster

HOW A JEWISH DRIVER, AN AMERICAN HEIRESS,

AND A LEGENDARY CAR BEAT HITLER'S BEST

Neal Bascomb

Houghton Mifflin Harcourt

BOSTON   NEW YORK

2020

For information about permission to reproduce selections from this book, write
to trade.permissions@hmhco.com or to Permissions, Houghton Mifflin Harcourt
Publishing Company, 3 Park Avenue, 19th Floor, New York, New York 10016.

hmhbooks.com

Library of Congress Cataloging-in-Publication Data
Names: Bascomb, Neal, author.
Title: Faster : how a Jewish driver, an American heiress, and a legendary car beat Hitler's best /
Neal Bascomb.
Description: Boston : Houghton Mifflin Harcourt, [2020] |
Includes bibliographical references and index.
Identifiers: LCCN 2019033972 (print) | LCCN 2019033973 (ebook) |
ISBN 9781328489876 (hardcover) | ISBN 9781328489838 (ebook)
Subjects: LCSH: Automobile racing—History. | Automobile racing drivers—Biography. |
Discrimination in sports. | Grand Prix racing—History. |
World War, 1939–1945—Social aspects.
Classification: LCC GV1029.15 .B347 2020 (print) | LCC GV1029.15 (ebook) |
DDC 796.7209—dc23
LC record available at https://lccn.loc.gov/2019033972
LC ebook record available at https://lccn.loc.gov/2019033973

Book design by Margaret Rosewitz

Printed in the United States of America
DOC 10 9 8 7 6 5 4 3 2 1

Maps by Lucidity Information Design, LLC

For Charlotte and Julia,

  May you embrace life's fearful joys—

I see you stand like greyhounds in the slips,
Straining upon the start.

—William Shakespeare, *Henry V*

Contents

Author's Note

A TERRIFIC ADVENTURE AWAITS, but I must hurry.
Midmorning, heading north on the I-405 from Los Angeles International Airport, I am stuck in a traffic jam. A sea of vehicles of every make surround me: long-haul semis, boxy sedans, Denalis with tinted windows, Priuses with Uber stickers, black town cars, landscaping trucks, and the occasional zesty convertible. My own rented black GMC Terrain is one of those nondescript compact SUVs that automakers stamp out with all the cookie-cutter variation of a Ford Model T. If parked in a crowded lot, I would fail to pick out my rental without clicking its key fob to trigger the lights.

None of us is getting anywhere fast. Ten minutes pass at a standstill. Then twenty. According to Google Maps, I have another fifty-eight miles — or one hour, fifty-two minutes — to go until I reach Oxnard. The line of my route on the screen map looks an ugly red. Surely they will wait for me before they ship the Delahaye 145 race car off to London to sit behind a velvet rope in the Victoria & Albert Museum. I try not to pound the steering wheel in frustration.

The bottleneck ahead finally loosens. Once I veer off onto the I-10 toward Santa Monica, the stop-and-go traffic becomes mostly go. Then I am cruising north on the Pacific Coast Highway. In the distance there are mostly vistas of ocean and wildflower-covered hillsides. I might make it after all.

My GMC is comfortable, but unexciting—a rental car obligation. Retractable seats with good lumbar support. Automatic transmission. Apple CarPlay to listen to my latest Spotify favorite, the Lumineers. Past Malibu, I battle with the electronic windows, unable to convince them to remain half open. Instead I seal myself into the air-conditioned cocoon. No salty breezes for me. At stoplights, the engine shuts off to save gas. It does not ask me for permission.

Halfway there I take a call from my wife in Seattle. She cannot even tell I'm driving. Whatever churns underneath the hood, it is quiet, reasonable, and unflappable, all worthy qualities in a vehicle meant to get one safely from point A to point B.

A very different car is being readied for me in Oxnard. It is a reward —and capstone—after two years of investigation into its long-forgotten history.

There was a period, shortly before the outbreak of World War II, when the Delahaye 145 was one of the most noted Grand Prix race cars in the world. Its exploits had equal billing to news stories of peace coming undone in Europe. Huge crowds assembled to watch it compete or to glimpse its shiny V12 engine up close at motor shows. When it first appeared at Montlhéry, an autodrome in France, many thought its design peculiar. One critic likened it to a "praying mantis" rather than a machine built for speed. After the Delahaye smashed records on the closed oval circuit, sought the "Million Franc" prize, and dared to be "The Car That Beat Hitler," the naysayers became adoring admirers. Its owner, its designer and builder, and the driver who often risked death pushing the Delahaye to the limit were heralded as national heroes.

Its story began in 1933, when the leader of the new Third Reich made reigning over the Grand Prix one of his first missions. His Silver Arrow race cars, piloted by the ruthlessly indefatigable champion Rudi Caracciola and the blond-haired, blue-eyed poster boy Bernd Rosemeyer, stood for more than sporting prowess: they represented the master race conquering the rest of the world. "A Mercedes-Benz victory is a German victory," heralded the Nazi propaganda machine. Hitler aimed to use their success to inspire hundreds of thousands of young men to enlist in the ranks of a motorized army, which its automobile firms, now transitioning into massive industrial machines, would help bring into being.

After years of unchecked German triumph, a woman called Lucy O'Reilly Schell decided that something must be done, so she launched her own Grand Prix racing team. A dazzlingly fine driver in her own right and the only child of a well-heeled American entrepreneur, she had cash to spend, reasons of her own to challenge the Germans, and the will to claim her place in a world dominated by men. For a car, she chose the most unlikely of manufacturers: Delahaye. Managed by Charles Weiffenbach, the old French firm was known for producing sturdy, staid vehicles, mostly trucks. Racing was a path to save the small company. For a driver, Lucy recruited René Dreyfus; once a meteoric up-and-comer, he had been excluded from competing on the best teams, with the best cars, because of his Jewish heritage. Triumph over the Nazis promised redemption for all of them.

My journey to uncover this tale of a team of strivers took me in many directions. Unlike the prewar stories of such sports giants as Jesse Owens, Joe Louis, or that "crooked-leg racehorse named Seabiscuit," time had largely erased from memory the endeavors of Lucy, René, Charles, and the Delahaye 145. No writer had devoted a book solely to the subject. At first, I sank myself into libraries like the marvelous Revs Institute in Naples, Florida. This was my first foray into automotive history, and the sheer volume of contemporaneous journals and magazines devoted to the Grand Prix—in multiple languages—stunned me. Day after day, I thumbed through thousands of pages, rooting out pieces of the saga. The Daimler-Benz Archives in Stuttgart and the National Library of France in Paris proved equally helpful.

I quickly learned that information on classic cars—and their drivers—is deeply cherished but often held closely by clubs and private collectors. Diligence and a little bit of persuasion gained me access to scholarly treasures held inside a sprawling French farmhouse, a cluttered Seattle garage, and a storybook English manor, among other places. The family of René Dreyfus shed a lot of light as well.

It is one thing to study detailed course maps of the significant races in this history. It is another to walk or drive them. La Turbie outside Nice. The Nürburgring in the Eifel Mountains. Montlhéry south of Paris. Monaco through the streets of Monte Carlo. Pau on the edge of the Pyrenees separating Spain and France. I wanted to know every hairpin, every straight, every rise and fall.

Throughout my research, I visited numerous car museums across America and Europe, wandering among the collections of Alfa Romeos, Bugattis, Maseratis, Delahayes, Talbots, Ferraris, Mercedes, Peugeots, and Fords. Polished to a sheen, these cars looked like works of art. And they were immobile, surrounded by walls, their purpose — to go fast — thwarted now.

I wanted — and needed — to experience a sliver of what the heroes of this story had experienced in the cars they raced, most of all the Delahaye. In response to repeated requests, Richard Adatto, a board member and curator of sorts for the Mullin Automotive Museum, invited me for a drive. The museum, in Oxnard, California, was founded by American multimillionaire and collector Peter Mullin, who owned all but one of the four 145s.

I finally reach the museum parking lot. I turn into it and come to a quick stop. Just in time. Someone calls my name as I climb out of my GMC. It is Richard, waving me over. Richard has a measured, no-nonsense demeanor that only occasionally is broken by an impish smile. In his sixties, and a builder by profession, he is also an expert on prewar French cars. He leads me over to a boxy white warehouse opposite the museum.

The tapered tail of a car sticks halfway out of the warehouse door. The Delahaye 145. The many black-and-white photographs I have seen of the race car do it little justice. Painted a sky blue, the two-seater stands long, lean, and low, appearing altogether like a tiger crouched in anticipation of a leap. It is over eighty years old yet somehow looks futuristic, particularly with the swooping lines of its mudguards hovering over tires that look too narrow to handle much speed.

After opening the toy-sized door, Richard folds himself into the driver's seat. Nate, the museum's cheerful mechanic, leads him through the ignition process. Pull this. Turn that. Press this. The engine fires up, smoke billows from the twin exhausts, and the thunder of the engine nearly deafens me. Several years have passed since the 4.5-liter V12 engine was last started, yet after Nate performed some routine checks and replaced its twenty-four spark plugs before my arrival, the Delahaye rumbled awake with barely any hesitation.

Richard and Nate take the Delahaye on a short round of the parking lot to make sure everything is okay; then the car is winched inside a trailer and secured into place. In a separate vehicle, Richard and I follow the trailer away from the museum and, after fifteen minutes, deep into the lemon groves outside Oxnard. The empty roads that thread through the groves are the perfect place to let the Delahaye run free. It does not like to stop and start, nor to putter away in traffic. It is a race car after all.

The trailer halts on the gravel shoulder of a side road, and down the ramp comes the Delahaye. Richard is first into the open cockpit, shoehorning himself into the driver's seat on the right side of the car. Nate opens the door on the passenger side for me. Despite my many entreaties to drive the Delahaye myself, I was denied — and not without reason. The 145 was worth many millions of dollars, and Peter Mullin did not need an uninsured amateur seizing up its engine or pitching it into a tree. Anyway, given how tightly Richard and I are pressed together in the narrow two-seater, there is little distinction between driver and passenger.

Nate hitches a belt around my waist, a likely useless safety feature that René Dreyfus did not benefit from during his days piloting the Delahaye. Then, as today, one drove with neither a roll bar nor a crash helmet.

The engine growls at idle, uneasy. Looming over the Santa Monica Mountains, a bright sun shines down on us. The lemons hanging from the trees look as big as grapefruit. The smell of oil pervades the air. My left hand grips the cold steel of a handle secured on the dash. My senses feel sharper than usual.

Richard steps on the clutch and shifts into first gear. We roll off the gravel onto the road and make a 180-degree turn to point in the direction of a long straight. At five feet, seven inches, I am the same height that René was. My eyes barely rise over the long hood. There is no windshield. I am seated low and at a slight angle, my legs stretched out in the footwell. For some reason, I can't shake the impression that I am riding in a mechanized sled, such is my body position and my proximity to the ground. When I reach out over the door, my fingertips nearly brush the asphalt.

Without a hint of warning, Richard vaults down the road, engine wailing as he shifts from first to second, then third. We move faster and faster, the wind sweeping back my hair. I turn to Richard. He holds the big steering wheel at ten and two, making slight, but constant, adjustments. There is that impish smile. He is loving this.

I am scared. My grip tightens on the handle. The Delahaye does not feel stable. It fights to hold a straight line. A deep ditch borders the road. A plunge there would surely be the end of things. I have a family. Young children. Ahead approaches a sharp turn.

Richard neither brakes nor eases on the throttle. My feet press pedals that do not exist. Richard swings the wheel counterclockwise, and the tires clip the gravel edge of the road as we enter the left turn. The Delahaye hugs tight to the ground as we make the turn. Coming out of it, Richard presses on the gas, then shifts gears again. The speedometer needle swings sharply. We are now devouring an uphill climb. The engine pitches to a high scream. Quickly we enter another turn, this one to the right. Again the Delahaye clings to the road. We enter a long undulating straight.

A quick upshift. The Delahaye jolts ahead, faster than before, past the rows of lemon trees. Any fear fades away. The wind presses my hair back. It ripples my cheeks. The acceleration forces my upper body against the seat. The engine rises to an ear-splitting howl and throbs all around me, alive. We rocket forward. I feel every dip and hump in the road, but am not jarred; it is like I am welded with the Delahaye. It is the same with every shift of the gears, every tap on the brakes. Time evaporates. The world distills into the band of pavement ahead and the surrounding rush of wind and noise.

"Remarkable," I whisper. "Remarkable."

We summit a small hill, and it almost feels like we are flying.

All of a sudden, Richard slows. We have come out of the groves and have reached a highway intersection. Trucks shudder past. From my perspective inside the low-slung Delahaye, they look like giants. With a break in the traffic, Richard turns onto the highway. After some grinding of the ancient gears, he quickly hits fourth. We race even faster than before, engine yawping, as we almost leap onto the tail of a boxy sedan ahead. The acceleration is incredible. Then a quick jab of the steering wheel, and we head back into the groves. A farmer among the rows stares at the Delahaye, mouth agape. Another lightning straight, and we return to where we started beside the trailer.

Richard cuts the engine. The Delahaye stills. I perform some yoga moves to unfold myself from the seat and stand outside the car. I feel the ground stable under my feet, rather like when you step off a boat onto land.

Richard tells me that we barely broke 75 mph. I am stunned, not only because it felt like we were moving much faster but also because that is almost half the speed René Dreyfus would have run the Delahaye during the Grand Prix. Half.

A bicyclist pulls up beside us and stares at the car, not quite sure what to make of it. He asks Richard and Nate a bunch of questions. He is truly spellbound.

While they chatter, I climb into the driver's seat of the motionless Delahaye. I grip the steering wheel and the gearshift. I rest my feet on the clutch and the accelerator. For a moment, I am racing through the lemon groves again, the sun hot on my face, the wind screaming in my ears. More than ever, I appreciate how remarkable a car the 145 was in its glory days, the force of will that Lucy Schell showed in seeing it built, and the incredible skill and guts that René must have had to pilot it against the titans funded by the Third Reich.

Enzo Ferrari called racing "this life of fearful joys." I never quite understood his words until that afternoon.

March 2019

Prologue

"We Will Write the History Now"

THE BEAST, LONG lurking in plain sight while the Allies stood
idle, pounced at last. On May 10, 1940, wave after wave of German bombers, their supercharged engines in high pitch, swept across
the dawn sky while armored columns rumbled overland. Into Belgium,
Holland, and Luxembourg the Nazis advanced, shattering the morning
quiet. Their paratroopers severed communication lines and captured essential bridges. Commandos dropped from glider planes and seized critical fortresses before they could stall any advance. In short order, panzer
divisions barreled deep into foreign territory. When French and British
forces hurried northeastward to Belgium to stem the attack, they fell
straight into the trap of expectations entrenched from the First World
War.

To their east, the main thrust of the German juggernaut charged
through the seventy-mile stretch of the Ardennes, forested hills once
considered as impenetrable as the concrete fortifications of the Maginot
Line that ran along the border between France and Germany. Within
days, the Nazi spearhead, supported by artillery barrages and aerial attacks, crossed the Meuse River in France, forcing the Allies to retreat.
By May 15, the French prime minister, Paul Reynaud, despaired over
the telephone to his British counterpart, Winston Churchill, that the
war was all but lost.

The French had some fight left in them, but it was at best panicked going up against what one witness called "a cruel machine in perfect condition, organized, disciplined, all-powerful."

At the news of the Germans' rapid advance, Parisians took flight, particularly from the toniest quarters of the city. Railway stations were crowded with passengers desperate for tickets on sold-out trains, while overstuffed cars and buses jammed the roads leading south from the city. At the same time, forlorn refugees from Belgium poured into Paris from the north. "With bicycles and bundles and battered suitcases, holding twisted birdcages, and dogs in stiff arms," observed *Life* magazine, "they came and came and came."

Fearing an invasion for more than a year, the French had safeguarded many of their finest treasures. In Paris, monuments were sandbagged, and the stained-glass windows of Sainte-Chapelle had been removed. Curators at the Louvre denuded its walls of masterpieces such as the *Mona Lisa* and its floors of priceless sculptures. Convoys of nondescript trucks hauled these artworks to chateaus across the country. Likewise, French physicists evacuated their supplies of heavy water and uranium, instrumental to the pursuit of a nuclear bomb. Priceless art and rare substances were not the only items squirreled away as the German blitzkrieg threatened Paris. Across the city, people stashed family heirlooms in cellars and buried them wrapped in oilcloth. One Parisian hid a batch of diamonds in a jar of congealed lard that he left on his pantry shelf.

In the Delahaye factory on the rue du Banquier in the working-class heart of the city stood four 145s. The manufacturer's production chief intended to see his creations secured away, whether by dismantling them into parts, hiding them in caves outside the city, or, like those diamonds in the lard, masking them in the open, their engines and chassis covered up with new bodies—or none at all—and their true provenance concealed. These masterpieces could not be lost in the rage of war, nor found by the Nazis. There was little doubt that Hitler wanted them seized and destroyed.

In late May, the Germans drove back the Allied forces into northern France, where they were forced to evacuate the continent at Dunkirk. Then the invading army wheeled toward Paris. Reynaud exhorted his countrymen to fight to the death to hold the Somme, while his feckless war committee debated where to move the government when Paris fell. His staff collected secret papers to be sunk in barges in the Seine or burned in ministry yards.

While the police were armed with rifles to thwart any fifth-column attack and an antiaircraft gun was placed atop the Arc de Triomphe, many Parisians maintained an oblivious calm. Then, on June 3, the Luftwaffe hit Paris. Likened by one child to a "swarm of bees," Stuka planes dropped over a thousand bombs, targeting most intensely the Renault and Citroën factories in western Paris, which had transitioned to war production, much as their German counterparts, most notably Daimler-Benz and Auto Union, had done years before. The attack killed 254 and wounded triple that number.

The exodus from Paris accelerated.

Two days later, the Germans launched the second half of their campaign to take France. At the Somme, they ruptured the French line, their panzer divisions overpowering the courageous but doomed army. The door to Paris was ajar, and Reynaud and his government abandoned the capital.

Onward the Wehrmacht pressed.

In the capital, the growing numbers of routed French soldiers with unkempt beards and muddied uniforms portended the inevitable. Finally, on June 14, motorized columns of the German army—including heavy trucks, armored vehicles, motorcycles with sidecars, and tanks—entered an undefended city. Soldiers clad in gray and green followed on foot. The streets were so empty before them that at one intersection a herd of untethered cows aimlessly wandered past.

The Germans fortified positions at key arteries across the city, but there was no reason for such caution. Residents were helpless to launch a revolt when their armies had already retreated to the south. Instead, from windows and half-open doorways, they gaped at the rows of Germans marching past in their heavy boots.

By the afternoon, swastikas flew from the Arc de Triomphe and the ministry of foreign affairs. An enormous banner was strung to the Eiffel Tower that read, in block letters, "DEUTSCHLAND SIEGT AN ALLEN FRONTEN" (GERMANY IS EVERYWHERE VICTORIOUS). Trucks fitted with loudspeakers threaded throughout the city streets, demanding obedience and warning that any hostile act against the Third Reich's troops would be punishable by execution.

Motorized German troops occupy Paris, 1940

On June 18, General Charles de Gaulle broadcast his own message to his countrymen from his offices in exile at the BBC in London. "Is the last word said? Has all hope gone? Is the defeat definitive? No. Believe me, I tell you that nothing is lost for France. One day — victory . . . Whatever happens, the flame of the French resistance must not die and will not die."

Marshal Philippe Pétain, the newly installed French prime minister, maintained the opposite conviction. He pleaded for surrender, and on June 21, Hitler rolled into the Forest of Compiègne in an oversized Mercedes to deliver his demands. Surrounded by his highest officials, including General Walther von Brauchitsch, commander of all German forces, Hitler emerged from his car. Never one to shy from symbols, he forced the French to sign the terms of capitulation in the same train carriage in the same clearing where the Kaiser's emissaries had surrendered on November 11, 1918.

Fifty miles away in Paris, the Germans solidified their control of the capital, targeted its Jewish population, and began expropriating whatever they wanted. "They knew where everything was," was the common refrain: the best hotels, the finest galleries, the richest houses, and even the most popular bordellos.

On the Place de la Concorde, the German army commandeered the famously elegant Hôtel de Crillon and its neighboring colonnaded mansion, which was owned by the Automobile Club de France (the ACF). Founded in 1895, and the first such club of its kind, the club organized the French Grand Prix. Its membership included some of the wealthiest, most influential men in the city. Spread out over 100,000 square feet in a pair of buildings constructed during the reign of Louis XV, the club's quarters were well suited to its prestige.

One day early in the occupation, a Gestapo officer accompanied by several subordinates strode through the arched entryway of the ACF. The club's mahogany-paneled bars, its private bedrooms, and its shaded terraces were of no interest to him. Neither was he there to dine in one of its chandeliered, gold-trimmed restaurants, nor to swim in its palatial pool surrounded with statues like a Roman bath. Instead, the officer headed straight to the library, a cavernous, book-filled space that also held the ACF archives and records of every race held in the country since 1895. They were an invaluable and unique resource, chronicling remarkable French wins and ignoble defeats alike.

"Bring me all the race files," the Nazi ordered the young ACF librarian. The voluminous records were boxed up and brought out on a cart. While his subordinates hauled them away, the Gestapo officer turned to the librarian. "Go home and never return here, or you'll be arrested. We will write the history now."

The tale of René Dreyfus, his odd little Delahaye race car, and their champion Lucy Schell was one of the stories that Hitler would have liked struck from the books. This is its telling.

Part I

A young René Dreyfus

1

The Look

MAY 19, 1932, another Thursday night at the Roxy bar in Berlin. The champagne flowed; a jazz singer crooned. Known for its celebrity patrons, the trendy night spot catered particularly to sporting types and their entourages. Heavyweight boxer Max Schmeling's crew nicknamed it the "Missing Persons Bureau": if Schmeling could not be found at home or in the ring, he was sure as hell at the Roxy.

René Dreyfus crossed the dimly lit bar in a well-cut suit. Of average height, with the wiry build of a horse jockey, he weighed no more than 140 pounds dripping wet. He sported a nonstop smile and brown eyes alight with what one veteran journalist labeled "The Look": "a stare of searing intensity and undying affection that lets you know, without a doubt, René was put on Earth to drive cars fast." The Frenchman was in Berlin for that exact purpose.

René drew up a chair at one of the round tables. His companions included fellow racers Rudi Caracciola, the foremost German champion; Hans Stuck, known as "The King of the Mountains" for his expertise in hill climbs; Sir Malcolm Campbell, who had recently set a land-speed record in his famed *Blue Bird;* Georg Christian, Prince of Lobkowicz, a well-to-do amateur from Czechoslovakia who hid his exploits on the track from his family; and Manfred von Brauchitsch, an independent Mercedes driver, also of noble blood, whose uncle Walther was a rising star in the German army.

Many likened Berlin at that moment in time to a modern Babylon for its unabashed, intellectual freedom, wild entertainment, creative life, and liberated sex. This exuberance only served to spackle over the fault lines underneath the Weimar Republic. Millions were jobless; the democratic government, led by the ossified General Paul von Hindenberg, was too paralyzed with dissension to help, and the Nazi Party was gathering strength. Its army of brownshirts were already attacking Jews, immigrants, and communists with lawless abandon. A French diplomat stationed in the capital described the mood in those days: "To tread its streets was to skim a quicksand."

At the Roxy, René felt on sure ground. Although Jewish by descent, he was not religious, nor did he give any thought to how his heritage defined him. As far as he knew, neither his background nor his faith mattered among his fellow drivers at the table. Whether Catholic, Jewish, Protestant, or atheist; high-born or low-; German, French, Italian, or Siamese — what mattered was how quick and sure he was on the race circuit week in, week out. In this alone René believed he was measured, and the politics of the country in which he raced was something he skimmed through in the newspapers on his way to the sports section.

While the six men smoked cigarettes, drank, and talked of the hotly contested AVUS race that Sunday, a striking, dark-eyed individual approached their table. He was Erick Jan Hanussen, the clairvoyant whose show at the Scala Theater sold out nightly. Brauchitsch knew everybody who was anybody in Berlin and invited Hanussen to sit with them. Not a minute passed before the drivers wanted to know if he had any insights into Sunday's race. Facing death almost on a weekly basis, they tended to be a superstitious bunch.

Hanussen eyed each of them closely before scribbling a couple of words on a slip of paper. He put the paper in an envelope, went over to the bar and handed it to the bartender, then returned to their table. In a somber voice, he advised them not to retrieve the envelope until after the race. "The victor is sitting at this table," he said, "but one of you must die. The two names are inside that envelope." With these chilling words, he walked away.

René and the others debated whether he was a charlatan, but soon the conversation turned to other topics and laughter returned to the table. All the while, the other patrons at the Roxy gazed at these matadors of the modern age, who seemed to exist in another world altogether, one infused with glory and devoid of fear. It was a world that René had sought to join since his earliest days.

René circled his Bugatti Brescia with its horseshoe-shaped grille. One last check. Only twenty years old on February 25, 1926, he needed his mother's written permission to participate in the race, the La Turbie hill climb, whose start was at the edge of his hometown of Nice, France. Given his boyish face, it was a sure thing that his ID would be checked by the officials. An oversized cap, loose slacks, and suit jacket added to the impression of a young man lost on his way to a neighborhood dance.

The young Niçois squeezed himself into the wicker bucket seat he had installed with the help of his grandfather. They did not need to remove the vehicle's fenders or headlights since René had entered the sports car category to have a better chance at winning a prize among the seventy-five competitors.

Preparing to crank the engine into life was his brother Maurice, the Brescia's co-owner and René's ride-along mechanic. A year older and a few inches taller than René, Maurice had none of the devil-may-care spirit of his sibling. It was the very reason Maurice insisted on being present. To win, René would risk somersaulting off the edge of a precipice, but not with his older brother aboard to pay the price as well. Keeping an eye on the engine's gauges was of secondary importance.

By the starting line, their mother, Clelia, and younger sister, Suzanne, were cheering them on. René turned the fuel-line petcock (shut-off valve), then pressured the tank with four strokes of the fuel pump to his left. *Ready,* he nodded to Maurice. A jerk of the crank, and the Brescia's four-cylinder, 1.5-liter engine jarred awake with a *blaaaaaatttt . . . blaaaaattt . . . blaaaaaatt.* As legendary motorsport writer Ken Purdy described it, "A racing Bugatti engine in good shape always sounds as if it were about to fly to pieces." Maurice climbed into the open cockpit beside René as blue oil smoke sputtered from the exhaust pipe.

The race officials waved René toward the slightly inclined start position. The Brescia stood alone, with blocks of wood placed behind the rear tires to keep it from rolling backward when he let go of the clutch. Overhead was a cloudless blue sky.

Pressing down on the accelerator, René roused the reverberating engine into a throaty *rrrrrraap . . . rrrrrraaaappp . . . rrrrraaaaaaapppppp*. More blue smoke cast a pall over the road behind him. Staring at the starter flag, he curled his fingers on the steering wheel and angled his body forward. A nervous shiver ran through his body. Maurice braced himself by gripping the windshield frame with his hand. Perhaps the other hand rubbed the rabbit's foot he always carried in his pocket.

There was no need for words between the brothers. Getting away quick and fast was everything.

The starter raised the flag. Any moment now.

The race would take only minutes. René needed to be at his peak. Tight turns. All out in the straights. It would be over before his nerves calmed.

The flag snapped downward.

René punched the accelerator as he released the clutch. The Brescia surged ahead, upward, toward the first sharp turn.

Considered the father of all hill climbs, La Turbie was first won in 1897 by the tire manufacturer André Michelin in his steam-driven, two-ton De Dion-Bouton at an average speed of 19 mph. The course ran along the Grande Corniche, a road built by Napoleon that climbed like a serpent through the mountains to an altitude of 1,775 feet. This perilous route, famous for its severe gradients, overhanging cliffs, jaw-dropping passages over ravines, and acute twists, was first cut by the Romans, and it had so many turns that even an arithmomaniac would have abandoned count. The course ended at the charming hilltop village of La Turbie.

The higher René pushed the Brescia into the hills, the more the course zigzagged. The low stone barrier along the cliffside offered slight comfort. Rounding out of the bend at Mont Gros, surmounted by the domed Nice observatory, he accelerated. If Maurice bellowed at René to slow down, he never heard him over the engine's dolorous bark.

René was alone against the clock. There were no cars ahead of him or behind him against which to measure himself. Speed was everything. The best time won, and he had only one shot. He must race at the extreme, testing his mettle, and the Brescia's too. Hundredths of a second could determine the winner.

Going into another hairpin, he shifted down into second gear. Despite bracing their legs against the sides of the cockpit, he and Maurice were sandwiched together. Then, coming out of the turn, he accelerated quickly, the *rrraaaappppp-rrrrrraapppppppppp* of the Brescia echoing across the jagged mountain flanks. He had a couple of hundred feet until the next bend. There was no sense in saving his engine or worrying about tire wear. The race was too short. He devoted every shred of concentration to fighting gravity and the unassailable advance of the clock.

René had driven the course countless times. He knew the sequence of every bend, hollow, turn, and rise, like a detailed topographical map imprinted on his mind. He knew precisely where to angle the front wheels into a corner; he knew how long to allow the Brescia to drift to the outside for the following turn; he knew the best gear at the precise moment for every point of the course.

There were moments when the Brescia looked like it might swerve into the stone bank or, worse, off a sheer drop. But René managed to emerge from each turn unscathed, sometimes using the handbrake on his right side to accentuate the slide so he was in the perfect place to accelerate when he came into the straight.

Finally, the course flattened. Ahead, a dense crowd of ladies in dresses and gentlemen in suits, children curled at their legs, lined the road. The finish.

René left nothing in the engine as he shot across the line, leaving a whirl of dust and bits of stone behind him.

A lifetime might as well have passed since he left the start. The clock: five minutes, 26.4 seconds, an average of 43 mph on the course. It was a very fast time, the fastest ever at La Turbie in his particular engine class.

Now he had to wait for the others.

By early afternoon, all the competitors had finished. Louis Chiron, "The Champion of the Riviera," established the best overall time, a second shy of breaking the five-minute barrier. René clocked the sixth-fastest of all competitors, beating his nearest rival in his sports-car class by almost a minute and a half. Rousing congratulations, a shiny medal, and a country dinner in the town of La Turbie were his rewards for his biggest win yet.

This was all very nice, but René was impatient to be a *professional* race car driver. It was an unlikely calling for the son of a middle-class Jewish merchant, but there was nothing more he wanted from life.

. . .

As a boy, René convinced Maurice to have their sister and her friend Lily join them on a ride down the town's longest hill in his two-seater pedal car. All would be fine, René promised. The four shot downward only to realize that the meager handbrake could not slow the weight of four kids moving at such velocity. To avoid plunging into the river at the end of the hill, René swung the wheel to the left, flipping the cart over. Afterward, Maurice threw up. René thought it only a thrill.

What René really wanted was to take command of his family's Clément-Bayard. On occasion, his father, Alfred, allowed him to stand between his knees and hold the steering wheel of the colossal touring car as they zipped around town. René craved a chance to drive before he thought to question why.

War against Germany in 1914 broke the idyllic spell of his childhood. Alfred was drafted into the French army. Soldiers requisitioned his Clément-Bayard. Within months, their mother, Clelia, and her three children had to flee their home in Mantes-la-Jolies, outside Paris, in advance of the Kaiser's troops. At Gare de Lyon, they tried to board a passenger train to Nice, but there were no available seats. Instead, they were directed to a line of cattle cars heading south. With passengers packed to its filthy walls, the train took over twenty-four hours to reach the coast. It moved so slowly that nine-year-old René often jumped from the car and walked alongside the train for fresh air. In Nice, then Vesoul, the young family waited for Alfred to return from the war.

Reunited after the peace, the family settled in Paris. Business blossomed again, now in raincoats and fine garments, but Alfred was unwell. On the front, he had suffered several German gas attacks, and his lungs never recovered. In 1923, the Dreyfus family returned to Nice, now a thriving cosmopolitan city, overlooking azure-blue waters, with all the bohemian culture of Paris but with better weather. René was eighteen. For fun, he went from racing bicycles to racing motorcycles. Cars were the natural next step.

Alfred urged his two sons to think toward the future, promising to set them up together in a business. René thought launching a movie theater would be glamorous: premieres, Sunday matinees, stars of the silver screen at his doorstep. Maurice wanted to buy a wholesale paper company. People might like the movies, but they needed paper. Within days of investing in the pragmatic choice for his sons, Alfred died of a heart attack.

René felt unmoored by the loss of his father and obligated to an enterprise he never wanted. The family sold their house and moved into an apartment. Their father's grand De Dion-Bouton V8 torpedo had already been replaced by a two-seater, 6-hp Mathis, an easier vehicle for René to drive around Nice and nearby towns to sell paper. Maurice remained in the shop to handle everything else.

On the sinuous, treacherous roads outside Nice, René schooled himself in how to drive fast — and loved the flood of euphoria he experienced being in tune with the Mathis as they careened through the hills. He and Maurice joined the Moto Club de Nice, home of what René called the "sporting young bloods" compared to the graybeards at the Automobile Club de Nice.

Maurice Dreyfus and his father, Alfred, in the Mathis, the first car
René ever raced

René entered his first race in 1924, forging his mother's name to do so. His eighty-year-old grandfather helped him fit a huge exhaust pipe to the Mathis and strip off the fenders. Maurice rode aboard as mechanic, always the protective brother. Uncontested in his category, René won the 750-cc class of the Circuit de Gattières. Clelia learned of his victory and forced her son to sell the Mathis for a sturdy Hotchkiss touring car. Undeterred, René competed in the Hotchkiss.

Conversation with his fellow Moto Club members and visits to local garages convinced René that he needed a Bugatti so that he could enter the better races. By one calculation, Bugattis won 1,045 races between 1925 and 1926. But the cars were as temperamental and distinctive as their creator.

Ettore Bugatti, better known as "Le Patron," was Italian by birth but had long ago moved to Molsheim, in northeastern France, to design and build automobiles. More artist than businessman, he was an inventive genius whose cars, including the little Brescia, had become the premier force in racing across Europe.

Before the Great War, while most car manufacturers were producing boxy-shaped behemoths whose big engines and overall weight were equated by their designers with better road handling, Bugatti thought the opposite. "Le poids, c'est l'ennemi [Weight is the enemy]," he said.

The Brescia, known in its original form as the Type 13, really came to the fore in the early 1920s. With a chassis only six and a half feet long, it was a "little box of speed" or a "giant killer" that weighed less than half a ton and could reach a maximum speed of 70 mph. Some praised its "silk smooth" four-cylinder engine and likened the car to a "marvel in the matter of weight disposition and suspension" that "held down to its course like a chalk mark on the road."

Others questioned its jerky clutch, clumsy steering, mushy cable-operated brakes, and a ride that could "jar the back teeth out of a clothing store dummy." An early driver stated it best: "I suspect that it's rather like flying a Sopwith Camel. In the hands of a truly skilled pilot it is the ultimate winning machine, but it could easily be the death of someone less competent . . . Perhaps, like a Camel, it should be kept for a special breed of lunatic."

René wanted the Brescia nonetheless. The Hotchkiss would simply no longer do, and like those early Sopwith Camel pilots, he believed himself untouchable. He pitched his brother—then his mother—to the idea that in the "nimble Brescia" he would be able to "get around faster and see more customers." Clelia did not buy what René was selling but agreed.

The Dreyfus family arrived at the Bugatti dealership in the center of Nice. It was run by Ernest Friderich, one of Le Patron's earliest champion drivers, who was happy to sell another Brescia to another upstart with glory in his eyes. It seemed that every young man in town wanted to become the next Georges Boillot, the two-time French Grand Prix champion, whose steely gaze and walrus mustache had been splashed across newspapers on a daily basis before he died a hero's death in a 1916 dogfight against seven German Fokkers. René was susceptible to the draw of becoming a national champion, but the truth was that for him racing meant having the freedom to do what he loved: drive fast and win races.

After his grandson's 1926 triumph at La Turbie, René's grandfather delicately clipped out any mentions of René from several Nice newspapers and glued them onto the first page of a large scrapbook, taking care to note the publication name and date of every clip in his fine script. Over the next few years, one page after the next in the scrapbook filled with clippings of René's wins or placements in hill climbs and provincial races across the Riviera. The shelves in his bedroom became crowded with trinkets, plaques, and trophies. He had yet to earn any income from his efforts, but he was now a dominating presence in the local events, particularly since Louis Chiron, his fellow Moto Club member, had graduated from the little leagues.

A half-dozen years René's senior, Louis had the looks of a film star, with the accompanying ego. In the early 1920s, while trying to make it as a race driver, he had earned his living as a dancer for hire at the Hôtel de Paris in Monaco. Wooing the wives of tycoons or minor royalty was his particular skill, and depending on who one asked, a Russian princess or a rich American widow had financed the first car that set him up in the sport. They called Louis "The Old Fox" for a reason.

He was a daring but calm driver, always "poised, handling his car as if it were made of glass, changing gear delicately, with two fingers." A series of win-place-or-shows caught the attention of Ettore Bugatti, who signed Chiron to his official factory team.

Every year, auto manufacturers like Bugatti, Alfa Romeo, Maserati, Fiat, Talbot, Delage, and Mercedes hired the best drivers for their stables. They earned salaries and splits of appearance money and prizes and piloted the most advanced cars, engineered by the sharpest minds and serviced by experienced crews. Team managers chose when and where the drivers competed, handled all the logistics, and made sure there was a spot on the starting grid for them.

There was always a race to be run. Every season, the Association Internationale des Automobile Clubs Reconnus (AIACR), the association of the various national automobile clubs of Europe, determined the formula that cars and races had to adhere to in order to be part of the official Grand Prix circuit. This formula could stipulate the length of the race, engine size or design, overall car weight or dimensions, types of fuel allowed, and other factors—all to enforce some degree of uniformity. Among the events, there were "Les Grandes Épreuves" like the French and Italian Grands Prix, which earned drivers points toward the European Championship, as well as two dozen other races.

Teams also participated in sports-car races in which vehicles had to be fitted with lights, fenders, and other road equipment that a driver would find on a typical highway. A win at some of these events, including the Targa Florio, 24 Hours of Le Mans, and Mille Miglia, were coveted the same as a Grand Prix triumph. Then there were the hill climbs like La Turbie and endurance contests like the Monte Carlo Rally.

To pursue his dream of joining a factory team, René abandoned the charade of working in the paper business with Maurice. The Bugatti dealer Friderich took René under his wing and managed his race schedule for a split of expenses and winnings. There were far more of the former than the latter. In 1929, Friderich secured René a spot in the inaugural Monaco Grand Prix, but the young driver was driving an outdated, underpowered Bugatti and barely ranked. Over the next year, he competed on an almost weekly basis, supported by his family and Friderich as he pushed for the race that would catapult him into the big time.

Monte Carlo was at its most alive after midnight. In casinos, under glittering chandeliers, gamblers eyed roulette balls sweeping around tracks and dice tumbling across fields of green baize. All about the cliffside town, revelers danced and drank in innumerable *boîtes de nuit,* while others toasted their good fortune under the twinkling lights of their sailing yachts, which rocked on the horseshoe bay.

It was April 2, 1930, and René Dreyfus should have been asleep—

he needed the rest. But he was awake in his hotel room, wrestling with doubts over his chances in the upcoming "Race of a Thousand Corners." In only its second year, the Monaco Grand Prix was already considered a premier event because of its sinuous 1.97-mile course through the hillside heart of Monte Carlo. It was the world's most glamorous "round-the-houses" circuit, and its 100 laps were a true testbed for drivers and their cars.

The month before, René and Friderich took the leap and purchased a 2.3-liter Type 35B, a three-year-old variation on the successful Bugatti race car, fitted with a supercharger. By drawing in compressed air and fuel from the carburetor, compressing the mixture, and blowing it into the engine's eight cylinders, the Bugatti's supercharger allowed more fuel to be pumped into each chamber, creating a bigger explosion when the mixture was ignited by the spark plugs and thus more power from the piston stroke. René entered two small races and won both. He felt like a "child with a new and better toy."

Le Patron had brought another variation on the model for his factory "works" team: the two-liter Type 35C. It was also supercharged but sported a higher-revving, shorter-stroke engine and a special axle ratio for better low-speed wheel torque, which would be helpful in a 100-lap race with hairpin bends, corkscrew descents, and uphill straights.

Before dawn, unable to sleep, René's mind sprang on an idea. His car had a 26.5-gallon tank, which would require refueling after sixty laps. If he fitted an extra tank, then he could skip the pits and save a couple of minutes. That could be the difference that meant a top finish.

Still in his pajamas, he hurried down the landing to his team manager's room. Groggy and irritated, Friderich answered the door. "Come on, we have some work to do," René said before rattling out his idea.

Friderich stared at his enthusiastic partner, two decades his junior. Finally, he said, "Good night. I'm going back to bed."

René pushed into the room. He only needed a nine-gallon tank.

"Where'll we put it?" Friderich asked, his bulldog jowls flapping. "On a trailer?"

"No, no. In the cockpit—empty seat—under the canvas—no one will see it." René never considered the risk of carrying a jostling tank of gas beside him as a passenger.

"It simply won't work," Friderich said. "That extra tank will just be deadweight. It will get in your way. You will have to stop anyway to clean your goggles, to get a drink of water — and get more gasoline."

No tank, no race, René threatened. None of the race's rules forbade it.

The two woke up Maurice to mediate their predawn standoff. Finally, Friderich said, "It's you who's driving," and the three men went straight to the garage to figure out how to make it work.

Early on the morning of the race, April 6, René drank a café au lait, ate a croissant, and then went down to the pits on the promenade. The dawn sun glittered off the Mediterranean. He and Friderich double-checked everything on his car, from the engine down to their special preparations for the second tank.

Afterward, René took a walk to calm his nerves. Monaco's small population had mushroomed in anticipation of the race, and spectators were gathered everywhere he looked: in the grandstands; on hotel terraces; aboard yachts, fishing vessels, and dinghies jamming the harbor; across the hillsides of "The Rock of Monaco," where the Prince's palace stood; and on rooftops and along every foot of the course, which had been fortified with low walls of sandbags. The whole town was a natural amphitheater. At certain points, onlookers could almost stretch out and graze the cars with their fingers, an immediacy René found intoxicating.

René returned to the pits and shared some cold chicken and Bordeaux with Maurice under the shade of a palm tree. Officials waved at the drivers to bring their cars to their places. The grid had been drawn by lots, and René was in the fourth row. There were seventeen competitors, and given the tight course, early positioning was critical.

This year the organizers had set up parimutuel betting on the race, and booths on the streets were already besieged by gamblers. It seemed that everybody, including the gendarmes, had a tip or a shred of gossip about the best driver or car. The hometown favorite Chiron paid two-to-one. A gamble on René earned three times that amount. René was badgered over the last few days of practice: "Think you've got a chance?"

As tradition had dictated since 1900 and the inaugural Gordon Bennett Cup, the cars were painted in colors based on their driver's or team's nationality — blue for French, white for German, red for Italian, green for British, and yellow for Belgian.

In his spotless overalls and crash helmet — the first worn in Europe — Louis passed René with a look, but no greeting. It was race day, and on such the Monégasque had no friends. He and the other drivers knew about the extra tank in René's car, but they scoffed that it would make no difference for the young independent.

René took one last check of his car, then bent down to ensure his shoelaces were triple-knotted. In a practice lap, a loop on his laces had caught on the clutch and brake pedal, and he had nearly jumped a curbstone into a wall. He shoehorned himself into the cockpit and took a sip from the long drinking straw sunk into his thermos of iced cola.

Friderich came to his side. "Don't force too much at the beginning. Wait until the twentieth lap. Then your engine will be warm." René nodded. Then he was alone. He adjusted his white cloth helmet and goggles and tried to settle down. Pandemonium surrounded him: the static squeal of the loudspeakers, the band playing, spectators stomping in the grandstands, the sudden churning of engines and coughs of smoke. He tried to force all of it from his mind.

At the line, the starter, Charles Faroux, raised a single finger, alerting the drivers that it was one minute until the start. The editor of *L'Auto*, France's preeminent sporting newspaper, and the founding director of the 24 Hours of Le Mans, Faroux was his country's doyen of motor racing. Wearing his trademark straw hat and an elegant suit, he held the red-and-white flag of Monaco behind his back, ready for the start.

From the sidelines, Maurice yelled good luck. René also spotted his mother and his sister Suzanne, who was eight months pregnant, and waved to them. Then, with a blur of sound and fury, the race began. The seventeen cars bolted wheel to wheel down the promenade toward the right-hand, cambered curve before the modest Sainte-Dévote Chapel. Coming out of the bend, René shifted into third to ascend Avenue de Monte Carlo. Louis was already in the lead. René was far in the back, but there was no room to pass in the mass of cars shooting up the 600-yard, inclined straight at 93 mph, looking altogether like a "multicolored serpent." René glanced at his tachometer — 5,300 rpm. *Hold back,* he thought. *Save the cold engine.*

At a slight bend in the climb, René punched down to first, his path impeded by a yellow Bugatti and a monolithic white Mercedes. Then, as he later recounted, it was back up "to second in the pullout, then

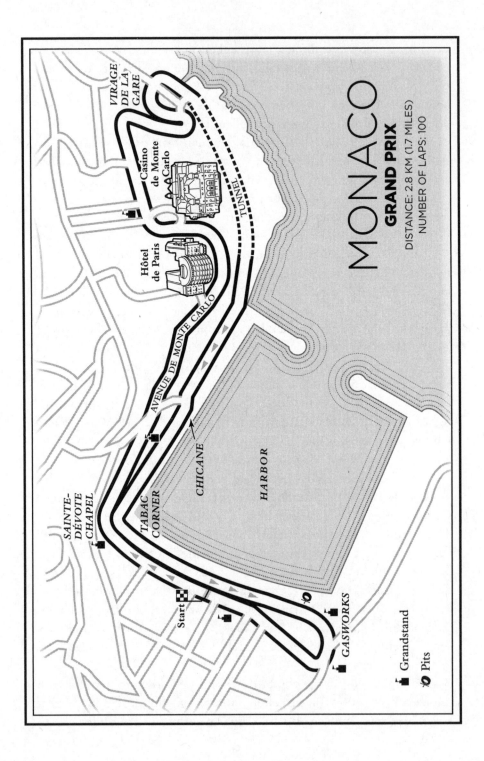

MONACO
GRAND PRIX
DISTANCE: 2.8 KM (1.7 MILES)
NUMBER OF LAPS: 100

VIRAGE DE LA GARE

Casino de Monte Carlo

Hôtel de Paris

AVENUE DE MONTE CARLO

TUNNEL

SAINTE-DÉVOTE CHAPEL

TABAC CORNER

CHICANE

HARBOR

Start

GASWORKS

Grandstand

Pits

third, wonder about fourth, then suddenly back down through the gears again." At the top, now high above the water on the city's cliffsides, he banked left alongside the Hôtel de Paris, then through the manicured gardens in front of the neo-Baroque confection that was the Casino de Monte Carlo. He tore down a short straight followed by three hairpin bends in succession—right—left—right—as he approached a "dive into a dull, stone-side ravine," as one writer put it, toward the seaside railway station. There were moments during those narrow turns when mere inches separated the nose of his Bugatti from the car in front. Another right curve at the water led into a 130-yard tunnel lit by flickering arc lights. As he accelerated through the soft long curve, the bellow of the engines in the arched stone tunnel reverberated inside his skull. Any errant tug at his wheel and his Bugatti would crash against the sides of the contained chute, tumble end over end, and his life would be over.

Emerging from the tunnel into the open again, he dashed down a short straight bordered with tamarisk trees. He remained at the tail end of the pack. A rapid left-right jog of the wheel brought him through the chicane. Then he burst along the harbor, swung a left at Tabac Corner, sped beside the water's edge, past the pits, turned a hairpin at the gasworks, and he was racing back toward the start.

Chiron and William Grover-Williams, the 1929 Monaco winner, jockeyed for first with almost record lap times, while the Italian Luigi Arcangeli nipped at their heels in his Maserati. René bided his time, moving up where he could. Positions changed so often that the leaderboard became a shifting, imprecise mess of names. The spectators— perched over balustrades, hanging out over terraces, climbing light poles —loved it. A *Motor* reporter regaled his readers: "What a crescendo of blaring exhausts, of hissing brakes, of the sharp *phut* as a cloud of blue smoke, redolent of castor oil, spurts from a supercharger, or the musical roar of a car's exhaust as the driver changes up through his gears. What an orgy of excitement."

In the tenth lap, René broke into the middle of the pack but was almost a minute behind the leader, Chiron, who was already lapping some competitors. Cars had begun to fall out of the race. Count Max Arco-Zinneberg's huge 7.1-liter Mercedes SSK smacked into a pile of sandbags. Mario Borzacchini's Maserati hit a wall. Others experienced brake-drum trouble or their engines seized. One German driver lamented that Monaco was "the Devil's course."

At twenty laps, René pushed into third place, but Chiron was a minute and a half ahead now. René kept to a fierce pace; sweat soaked his overalls. He drove the Bugatti like he was conducting a brutally swift symphony with his gloved hands. Streaking about the course, he shifted gears almost continuously, accelerating to 100 mph in fourth, then down to 10 mph in first. His grip tight on the wheel, he swung left and right, braked, shifted again. On average, there was a turn every twelve seconds. Throughout he kept watch on his fuel and oil gauges, switching petcocks and regulating the pressure with the hand pumps when needed. His arms ached, and his fingers and palms blistered and grew numb.

The blast of a cannon from a yacht in the harbor announced the halfway point. Only the durable Bugattis, ten in total, continued to run, but the crowd sensed it was a two-person contest between René and Louis. Still, Louis was almost a whole lap ahead.

With each round of the circuit, René nibbled several seconds from his rival's lead. Shortly after the sixtieth lap, his main tank was ebbing low, so he switched open the fuel line from the auxiliary tank on the seat beside him. Gasoline flowed smoothly to the engine. He would not have to stop in the pits. Louis would.

Into the seventieth lap, the Monégasque remained a minute and eighteen seconds ahead. There was a hustle of movement in his pit, a sure sign that he was about to refuel. Friderich waved his hands to René as he passed, signaling this information. Several laps followed, but Louis did not go into the pits. René could barely see through his oil-slicked goggles. Stopping to clean them, to rest his hands, or to gather his breath would cost precious seconds. Now, with the deadweight of the tank lightening, he could go even faster, go even more nimbly through the corners.

Finally, on lap eighty-three, Louis pulled into the pits. When René came around Tabac Corner into the harbor straight, he saw Maurice and Friderich on the sidelines jumping up and down. "He's there! He's there!" As he sped down the parallel straight, he spotted Louis climbing the Avenue de Monte Carlo, shoulders hunched, bent over the wheel, like he was willing himself forward. He had lost fifty seconds in the refueling. René was now ten seconds behind. Within reach. The chase was decidedly on, and he felt like he had sprouted wings.

Everyone among the spectators was asking the same question: "Could Dreyfus do it?" In reply, René tightened contact over the next two laps. Then he saw his opening in one of the hairpin turns in the descent to the tunnel. With a burst of speed, he seized his chance and, sliding out of the corner, claimed the lead. Three car lengths behind, Chiron ripped off his blue crash helmet and tossed it aside.

The fray continued. On the uphill straights, Louis tried to use his superior acceleration to regain the lead, but any gap he bridged was soon widened in the series of corners that followed. Their pace was blistering. On lap eighty-eight, René set an extraordinary average of 60 mph, reducing the record lap speed to two minutes, seven seconds. He pulled away by five seconds, then ten. Louis began to have trouble with a sticking accelerator. René increased the gap to twenty seconds and repeated to himself, "Be very careful now . . . you *cannot* make a mistake . . . you *are* going to win the Grand Prix of Monaco."

At the finish, Faroux waved the checkered flag, and a stampede from the pits encircled René. Friderich and Maurice embraced him at the same moment, almost knocking heads. Strangers clapped him on the shoulder and thanked him for their big payoffs in the parimutuel.

Louis was in a rage. At first he refused to look at René or to shake his hand. That was fine with René, whose hands were a blistered, throbbing mess. When he received the trophy from Prince Louis II of Monaco, he held it in the crook of his arm.

Many celebrations and dinners followed. René drank magnums of champagne, smoked cigars, and basked in the spotlight of his victory. *L'Auto* headlined, "In magnificent style, young Dreyfus in a Bugatti triumphs over Chiron." Newspapers featured his boyish face, smudged with oil, on their front pages. Between his first-place prize and sponsorship bonuses, he earned a generous sum of almost 200,000 francs (roughly $130,000 in today's dollars). More important, he had claimed his first major Grand Prix.

He followed with a victory at the Grand Prix de la Marne at Reims-Gueux, and then, sure of his prospects, traveled to Molsheim to ask Le Patron for a place on France's best factory team. Embittered over his

A weary, but victorious René Dreyfus and his sponsor,
Ernest Friderich (on the left), after the 1930 Monaco Grand Prix

team's defeat at Monaco, Bugatti refused to even see him. The snubbing
struck René almost physically.

He returned heartbroken to Nice, where a note awaited him from
Alfieri Maserati, one of six brothers involved in the young Italian automobile firm. Represented by the trident symbol, their race cars were
fast and sleek but unwieldly to pilot. One contemporary said that their
overly flexible chassis "jumped about on its own suspension like a cat on
hot bricks." Nonetheless, Maserati had a tremendous 1930 season, and
René hurried down to Bologna.

There, at the ramshackle Maserati factory, a lone secretary was peck-
ing away at her typewriter in a closet-sized office. She waved René in
to meet the brothers, who wore the same double-breasted blue overalls
as their workers. The Maseratis gave him a tour, and then they lunched
on an impeccable stew at their favorite restaurant, Il Pappagallo. René
signed on to be their lead driver for the following year, the shimmering
hope of his youth realized.

In his first race with Maserati at the 1931 Tunis Grand Prix, René suffered a terrible crash that accordioned his car. A feeling spawned that a dark cloud was following him. Still, he managed to race well enough in his initial season to be invited back for 1932. He agreed to return, less because his prospects of winning were good and more because he simply liked the brothers. They involved him in the preparation of their race cars. "What do you think about this?" "Let's discuss that," was their operating style. Together they went over blueprints. Together they adjusted details on their engines. Together they took test drives in the countryside, often an exercise in literally avoiding chickens crossing the road. There was no pretense. All the brothers were mechanics themselves. They were hard-working, honest, and serious. At the end of each day, they dined together, and René was beginning to jabber away in Italian, feeling part of a tight-knit family.

His ill fortune working with the family refused to lift in his second season. He suffered repeated engine problems and slammed into a house at the Monaco Grand Prix because of faulty brakes. At AVUS, Hanussen's eerie prediction in Berlin's Roxy bar hung over his every move.

May 22 was a stultifyingly hot Sunday afternoon for a race. Those Berliners not at AVUS had left for the cool shores of Lake Müggelsee or Lake Tegel. Still, a big crowd surrounded the 12-mile track. Children climbed trees to sit on the limbs for a better view. Many others broke the slats off the fence barriers for the same reason. The grandstands were filled with celebrities—film stars and crown princes here, flying aces, directors, and musicians over there.

René took off in his lipstick-red Maserati from the start. With its six-teen-cylinder, supercharged five-liter engine, the Maserati thundered like an aircraft barreling down a runway. It was the fastest car he had ever driven, and he led the race after the first lap. Close behind was Rudi Caracciola, in a 2.3-liter Alfa Romeo, and Brauchitsch, in his Mercedes SSKL (the "L" signifying a lighter version of its predecessor the SSK).

For 1932, the Grand Prix was essentially *formule libre:* there were no restrictions on engine size, weight, or fuel consumption. This allowed competitors to field cars of every sort, their common aim being to increase their vehicle's speed. Their handling had not received the same attention.

Prince Lobkowicz was driving his white-and-blue Bugatti T54, a car that was notoriously difficult to pilot. At the first turn, before the south loop, he found himself pressed between two racers. He gave way but when trying to avoid a clip of the grass to his right swung his wheel a fraction too sharply to the left. At 125 mph, the Bugatti sheared sideways, barreled across the grass strip separating the straights, then leaped upward, tumbling and flipping for some sixty feet until it struck a tree. It settled in a mangled wreck on a railway embankment. Lobkowicz's skull was fractured in the crash, and he never regained consciousness.

The race continued unabated. Early on, René set a new lap record at 130.5 mph, but then his accelerator began to stick, keeping his speed up even when he wanted to slow. He managed to release it before hurtling off the track. Then he turned into the pits, cut the engine, and pushed the Maserati toward his slot. "It's over," he said breathlessly to his team as he pulled up.

"No," Ernesto Maserati rebutted. "You must finish the race for your lap record to count." The mechanics and Ernesto inspected the car. Ernesto identified the problem, tore off a piece of tin from the signal board, and jammed it somewhere into the crevices of the Maserati's floorboard.

"It won't work," René said.

"Just finish the race," Ernesto pleaded. "If the accelerator gives you problems, there's always the button on the dash to stop the engine."

René returned to the track, his mind churning over what he would do if the accelerator stuck while he flashed toward the north or south loop. Change gears, hit the brakes, and punch that button.

The long pit stop and the accelerator trouble put René in dead last, where he stayed throughout the rest of the race, but he finished—and was alive. Brauchitsch, who had flailed about in his SSKL cockpit like he was caught in a tempest, narrowly won over Rudi Caracciola, who was like an automaton.

Later at the hotel, René went to see Ernesto. The faulty accelerator had almost got him killed. René wanted off the team, sure enough in his ability now among the elite drivers that he would easily find a new home—one with more reliable cars.

"Please release me," he pleaded.

Ernesto agreed.

That night, Brauchitsch rang the Roxy bar in Berlin to find out what was written on the piece of paper inside the envelope. The bartender read two names: "Lobkowicz. Brauchitsch." Word of the prediction quickly spread among the drivers.

Forces beyond their control seemed to be steering their fates, and René could not help but think that his name could have been in that envelope just as easily as Lobkowicz's. The dream of being a professional race car driver was inexorably bound to the risk of losing everything to maintain his place in the world. He was not alone in facing it.

2

The Rainmaster

I N A SNUB-NOSED Alfa Romeo, Rudi Caracciola shot around the fourteenth — and final — lap of the 1932 Eifel race. Through the steep rolling hills, hollows, and forests south of Bonn, he maintained his lead over his competitors on the Nürburgring, a track that one historian noted must have been built by "an intoxicated giant sent out to trace the road."

Only René Dreyfus was within contention, twenty seconds back. Driving one of Louis Chiron's privately owned Bugatti T51s — the more efficient and more powerful successor to the 35B — René was running a strong race, particularly considering that he had only left Maserati the week before and this was his first time battling the unforgiving "Ring."

Among the 120,000 fans who lined the 14-mile course that threaded around the medieval ruins of Schloss Nürburg, few, if any, thought that the Frenchman had a chance this late in the race. Rudi would need to make a mistake, and he rarely made mistakes, even on a course whose elevation changes were almost as frequent as his gearshifts. A smooth, rational, and imperturbable driver, he always coaxed the best out of whatever car he was driving but never overestimated its ability. "Rudolf Caracciola was the most deceptive driver," a contemporary wrote. "He never seemed in any hurry, and yet look at his record!"

As expected, Rudi won the checkered flag. The only surprise of the race was that he had piloted an Alfa Romeo instead of a Mercedes. For years he had been one of the Stuttgart firm's foremost stars, and his name and its brand were almost synonymous.

Rudi liked Mercedes, but he was devoted to whichever car gave him the best chance to win. Like many drivers, his allegiance to any one country or company, even the one that had secured his place in the sport, mattered little. He might never have had a chance to be so single-minded — or raced professionally at all — if not for a haphazard brawl that set him on his path in the first place.

The band at the Kakadu nightclub in Aachen, Germany, roused up the crowd to its feverish beat. Sitting in a red upholstered booth beside his coworkers, Rudi Caracciola drank a brandy and smoked. The twenty-two-year-old was not one for dancing, and anyway, he was tired, having just finished up his mechanic's shift at the Fafnir Auto Works.

Even though it was 1923, Aachen remained occupied by Belgian forces as part of the Versailles Treaty. They were a ubiquitous, and un-welcome, presence in the town.

An argument broke out between the group of mechanics and some Belgian officers. One of them, as wide as a door, pushed his way over to the booth through the crush of clubgoers. Rudi edged out of his seat.

As the Belgian officer wheeled back to hit one of his friends, Rudi sprang from the booth and struck him. The Belgian's nose cracked like a walnut, and he crumpled to the floor.

A second later, Rudi and his friend burst out of the door of the Ka-kadu and ran through the maze of cobblestone streets until they stopped under the shadow of the cathedral. As Rudi caught his breath, the im-port of what he had done fell like an anvil. Assaulting an occupying sol-ider was a serious offense.

"You've got to get away," urged his friend. "This very night."

A clatter of hobnailed boots sounded through the streets — a patrol, out already searching for him. The two mechanics hid as a band of Bel-gians passed, rifles slung over their shoulders. Aachen was a small city, and they would already have a description of Rudi. He wore a rumpled blue suit, one of more expensive quality than a mechanic could be ex-pected to afford. Just shy of six feet tall, he was built like a spade, square-shouldered and slim-waisted. He had dark hair swept cleanly back from a high forehead, over coal-colored eyes, a pug nose, and a square jaw.

Rudi always had an unfettered appetite for trouble, one that he first fostered at his family home, the riverside Hotel Furstenberg in Remagen, Germany. Before he was out of knee socks, he was known as "The King of the Rhine Valley Scoundrels," a suitable nickname for a boy with a noble bloodline that ran a thousand years back to roots in Italy, hence the non-Germanic name Caracciola. When Rudi was fourteen, he lost his father during World War I, and his mother Mathilde was too busy managing their hotel afterward to curb her son's growing rebellious streak. He worked the manually operated elevator like it was a vertical racetrack. He borrowed a guest's Mercedes before he could see over the dash—lurching ahead, grinding the gears, nearly hurtling off the road. He also ran a bootleg wine operation and commandeered the hotel's yacht to charge passengers to cross the Rhine. His mother chalked up the incidents to high-spirited youth and smoothed over the uproars that followed.

Most of these antics, including ordering thick volumes of automobile catalogs "for the hotel" on the sly, had a common theme: mastering the newfangled contraptions of speed. Rudi came of age during the era of the record-setting Blitzen Benz and the winning of the 1914 French Grand Prix by his countryman Christian Lautenschlager over Georges Boillot on the eve of war. Motor cars would always come first for him, before school and before running the hotel. His mother tried to convince him otherwise, but he did not bend.

The family thought he might come around if he actually had to work in a gritty automobile factory. The job they secured him at Fafnir only heightened his enthusiasm, and he even won a few chances to enter their cars in some races. While competing, Rudi found that the faster he drove, the calmer and more at ease he became. He struggled to find such peace in any other part of his life.

After a midnight motorcyle ride to escape Aachen, Rudi arrived back home and awakened his family. They gathered around the breakfast table. "Done something foolish, my boy?" his mother asked.

The family hotel was also in the occupied Rhineland, and they risked trouble if he stayed. His mother gave him 60,000 marks, and his sister Hertha handed him the business card of S. T. Rathmann, a manufacturer she had met on the train who was from Dresden — outside the Allied occupation zone. Rathmann might be able to offer her brother a job in the automobile business.

Rudi left on the next train east.

He had high hopes until he found himself on the doorstep of a bland gray tenement. The middle-aged Theodor Rathmann had nothing to do with cars; he produced wooden dolls. When Rudi explained his ambition to become a race car driver, Rathmann was unusually generous. He helped Rudi find a job as a car salesman and promised to see what he could do about the racing too. Over the next few months, Rudi managed to sell only a single automobile, while he quickly burned through the money his mother had given him. Nonetheless, he focused all his waking efforts on race cars. Rathmann and his Dresden cabal found him a little roadster to borrow for a stadium race. They rewarded his victory with a barrel of beer, then introduced him to Director Herzing of the Daimler automobile company. They produced the Mercedes race cars (and would merge with the Benz & Company in 1926). Fate had handed Rudi another gift.

Later that same year, Rudi crossed a muddy yard between two hulking warehouses with sooted glass roofs. The steady rain that fell did little to dissipate the smell of oil and welded metal. Rudi could see sparks fly on the other side of the wide sliding doors and hear the drumbeat of heavy machinery. He was in Untertürkheim, the epicenter of Daimler's factories alongside the Neckar River in northeast Stuttgart.

The office boy accompanying him stopped in front of a large wooden shed. In its doorway stood the unbodied chassis of a race car fitted with tires and a pair of rough-hewn wooden seats. Rudi felt his heart drop into his stomach. After much pleading in Director Herzing's office, he now had his chance to test-drive for Mercedes.

Out from the shed came a haggard figure in blue overalls. At thirty-two, Christian Werner was already part of the racing team's old guard. He looked at Rudi with a mirthless, uninterested gaze, then gestured for him to get behind the wheel. He took the passenger seat beside Rudi.

Rudi drove out through the factory gates. Werner gave brusque orders: Left. Right. Straight. Over there. Now here. Over the next half hour, Rudi showed off his skills on the sloped hills beside the Neckar River. Quick as he could in the straights. Tight as he could in the bends.

Rain poured into the open cockpit, and the rear tires spat the rainwater in a curved arc behind them.

Finally Werner pointed Rudi back toward the factory. He pulled up to a crisp stop by the shed. Werner bid him, "God bless," and headed toward the half-lit shed.

"And how was it?" Rudi asked.

"I'll tell them upstairs." Werner disappeared.

Hours seemed to pass before Herzing called Rudi back into his office and offered him a sales position.

"But I want to be a racing driver —"

"Don't be an idiot," Herzing snapped. "Racing is no profession. First be an employee, then one day perhaps you can drive as much as you want."

Back in Dresden, Rudi languished in the Mercedes showroom. Situated on the city's busiest, most fashionable street, the showroom easily attracted customers, but Rudi failed to make a single sale. He spent most of his time gazing at the young ladies who stayed at the lavish European Hotel catty-corner to the showroom. One in particular, a beauty with cropped hair, dark eyes, and fair skin, caught his attention. A frequent guest at the hotel, she always took the same second-floor room and, on occasion, stood at the window watching the passing traffic.

Rudi shared a few furtive glances with her, once even a smile, but she was rarely alone. The hotel concierge told Rudi her name — Charlotte Liemann — and that she went by "Charly." She was the daughter of a well-known Berlin restaurateur and married to her father's chief assistant. The fact that she was married failed to cool Rudi's ardor. One afternoon he attended a dance at the hotel when Charly was in town. He asked for her hand, expecting another slow foxtrot, but the band switched into a tango, and he found himself lost in the thrill of being close to her.

"You're not very talkative," she said.

"I'd like you to watch me racing someday," Rudi choked out through his dry mouth.

"Why?"

"Then you'd have more regard for me."

Charly chuckled, and the two left it at that.

Since starting at Mercedes, Rudi had badgered his bosses for a car to race, no matter the event. Finally, they lent him a touring car for a local race. A win brought more opportunities and faster cars. Between stints at the showroom in his suit and starched white shirt, he traveled with the Mercedes team and participated in smaller races while the lead drivers, Werner and Otto Merz, competed in the Grand Prix.

Often Rudi was little more than a hired hand, assigned to drive one of the trucks or a chassis in search of a body. Regardless, he got to watch the greats at work, met the company's technical director, Ferdinand Porsche, and joined postrace dinners. He spent most of his time with Alfred Neubauer, an Austrian driver several years his senior who shared his duties and hotel rooms on the road.

Rudi and Charly grew closer. Her marriage was cheerless, and since that first dance, she had simply found that she and Rudi fit together. She filed for divorce. Over the course of the next year, Rudi's prowess as a driver earned him even more victories and favor with Mercedes. Yet he was discontented. He wanted to compete in the fastest cars on the greatest stage of all: the Grand Prix, a series of races that debuted in 1906, inexorably tying the development of the automobile with sport.

The concept of a self-propelled vehicle had long intrigued inventors and scientists. Leonardo da Vinci sketched plans, as did Isaac Newton. As far back as the sixteenth century, horseless carriages fitted with sails had maneuvered the roads outside Amsterdam. In the 1770s, Nicolas-Joseph Cugnot, a French military engineer, built a steam-powered tricycle to haul artillery. Sixty years later, in England, a steam carriage rolled along a country road at 12 mph, its engine stoked by a boy with a shovel. Farmers blocked its path with trees to prevent it from spooking their livestock.

The four-stroke internal-combustion engine changed everything. It was an invention with many fathers, but it was mastered by Gottlieb Daimler and Nikolaus Otto in Germany in 1876, when they gave the world a small, lightweight, efficient machine that converted power into motion. In 1886, their countryman Karl Benz patented a gasoline-powered vehicle. It boasted a one-cylinder, water-cooled engine, three wheels, a tubular frame, brakes, and buggy seats. The automobile had arrived.

Dozens of manufacturers across Europe and the United States sprang up like weeds after a spring rain. Initially they produced a motley bunch of cars. The driver sat on an open-air bench, using a wobbly tiller to steer. The engine was housed in a high square box. Solid rubber covered wooden artillery wheels. A brake lever pressed a block of wood against the tires. A bulbous copper horn warded away pedestrians. The headlights were like lanterns.

The engines grew to two cylinders, then four, and beyond. An increase in speed followed. Pneumatic tires arrived as well as steering wheels, though at first they were big enough to helm a yacht. Drum brakes replaced the blocks of wood. The roofless vehicles were slunk lower and longer in shape. Mechanics were needed on board to operate the hand pumps and the tubes feeding lubricant oil and fuel to the huffing, puffing beast of an engine.

Maneuvering the first cars was no leisurely affair. It took strength, moxie, and, most of all, an ability to endure being "boiled from the waist down and froze from the waist up."

The French president Félix Faure tartly remarked at an early exhibition, "Your cars are quite ugly and smell quite bad." Regardless, automobiles sparked the public's imagination, and their popularity spread when drivers began competing in town-to-town road races. Hundreds of thousands attended the spectacles. First Paris to Rouen. Then Paris to Bordeaux, Marseille, Berlin, and farther afield. Before the races started, crowds mustered in the half-light of dawn around vehicles — De Dions, Panhards, Daimler Phoenixes, Peugeots, Delahayes, and Benzes — of every shape and size. The drivers wore heavy leather coats, goggles as big as snorkel masks, and thick gloves to protect their hands from turning to bloody pulp while steering these lumbering beasts. With a laconic statement from the organizer — "This is Paris; over there is Berlin. Get there as soon as you can" — the race would begin.

There were no road signs, few depots at which to refuel, and no decent maps. Drivers sped through clouds of heavy brown dust, navigating by the telegraph poles that lined the road. Knives were used to peel off burst tires. At night, vehicles were put in paddocks, and the oil-soaked drivers slept in hotels only if they were not carrying out repairs. One early competitor fixed his broken chassis with pieces of a dismantled wardrobe from his hotel room.

Competition fueled innovation. One of the big leaps came with a new model of car launched at the turn of the century. It was designed by Wilhelm Maybach, built by the Daimler company, and instigated by its Nice representative, Emil Jellinek, whose daughter was the car's namesake. The first Mercedes marked advances in every aspect of the automobile, from engine efficiency to suspension, chassis, steering, and brakes. The car managed speeds of just over 50 mph. Piloted by Wilhelm Werner, the Mercedes beat all comers at one of its first competitions: the 1901 La Turbie hill climb.

Year by year, the cars—and the races—grew bigger and faster. And deadlier. Just before dawn on May 24, 1903, 221 vehicles of every sort launched themselves down the dust-choked, spectator-ridden roads of Versailles in the race from Paris to Madrid. Some were mammoths with eleven-liter engines that roared at 90 mph along the straights, despite having neither the steering nor the brakes to handle such velocities.

One car flipped over and caught fire; another smashed into a pile of stones and exploded apart; when the steering column on another jammed, the car met a tree. As the afternoon passed, the number of smashes multiplied. Over a dozen drivers, mechanics, and spectators were killed, and there were too many injured to determine a casualty count with any accuracy.

The disaster might have cemented the idea that these race car drivers—a medley of eccentric engineers, rich thrill-seekers, hired hands, and entrepreneurs eager to advertise their wares—were simply mad. Instead, this early tragedy burnished the drivers with the shine of the heroic. "Motor racing is unique," noted Robert Daley, a *New York Times* racing correspondent. "There is no other sport so noisy, so violent. There is none so cruel." Those willing to dare it were increasingly draped in glory and riches.

The drivers themselves felt they were part of a vanguard mastering the element of speed. The risk of death may have been the price, but it was worth it for the exhilaration of teetering at the edge of the impossible, the landscape passing in a fogged blur, the thrum of the engine melding into one's heartbeat.

Tragedy did, however, prompt organizers, most notably the Automobile Club de France, to abandon open-road races for controlled circuits with guidelines that determined what kind of car could compete. In 1906, the ACF put together a race on a 64-mile triangular circuit protected by a palisade outside Le Mans, France. Only cars that weighed under a maximum of 1,000 kilograms were allowed to enter the twelve-lap, two-day event.

Hungarian Ferenc Szisz, driving a Renault on two consecutive summer days so infernally hot that his tires melted off their rims, and with an average speed of 63 mph, beat out thirty-one others to win a trophy and 45,000 francs in prize money. He also secured his place in history by winning the first ever Grand Prix race. Others, like Rudi Caracciola, would find even greater fame on this stage.

In lashing rain, on the south bend of the AVUS racetrack outside Berlin, on Friday, July 9, 1926, Rudi pulled his Mercedes up to a halt. Moments before, two cars had collided during practice for the inaugural German Grand Prix. The sight of the twisted wreckage sent a shudder through him as he climbed out of his seat to help. Two emergency workers bundled Italian driver Luigi Platé into an ambulance. Left on the ground was his mechanic, Enrico Piroli. Eyes still open, he lay flat on his back, his arms and legs spread wide. Rain pelted his motionless body and fell like tears from his cheeks. While Rudi stared in disbelief at his first dead body, a paramedic draped a sheet over Piroli, leaving only his white canvas shoes sticking out from under the cover.

The test runs canceled, Rudi returned to his Berlin hotel. The sight of those muddied white shoes haunted him.

A month earlier, Rudi had approached Max Sailer, a former driver and now head of the Mercedes race department, with a proposition. The Mercedes works team was competing in the Spanish Grand Prix instead of the German race because the company wanted to show off its cars to the export market. Rudi saw his chance. He told Sailer that he wanted to represent Mercedes at AVUS.

After being pestered for a couple of hours, Sailer relented—partly. He promised Rudi a car and a support team, but only if he took part as an independent. The company dared not risk a mediocre showing by someone from the official Mercedes team. Adolf Rosenberger, another skilled young driver, would similarly participate. The company's technical director, Ferdinand Porsche, delivered to the drivers a pair of cars with straight-eight engines that were powerful, but temperamental. As

MERCEDES-BENZ CLASSIC ARCHIVES

1926 German Grand Prix winner Rudi Caracciola
with his copilot Eugen Salzer

a Mercedes aficionado wrote, the model was designed "with the rather pious hope that the driver was equal to his task."

On the dark gray Sunday afternoon of the German Grand Prix, Rudi waited tensely at the line. Sitting beside him was his young mechanic, Eugen Salzer, ready to go. The field was thick with thirty-nine cars. Rudi's training runs at AVUS had gone poorly, and he continued to have visions of Piroli. Rosenberger and a half-dozen other drivers had clocked much faster times. Moments before the start, Rudi scanned the grandstands for Charly, hoping to meet her gaze, but she was hidden among the throng.

The starter whipped the flag downward, its snap lost in the growling thunder of revving engines.

Instead of joining the cacophony, the Mercedes engine stalled. "Quick, man," Rudi yelled at Salzer as his competitors sped off. "Give us a push!"

Salzer jumped from the cockpit and shoved the Mercedes from behind. Rudi released the clutch, and the engine kicked into life. Salzer leaped back inside just before Rudi took off. Over a minute had passed.

Rudi tore down the 6-mile straight at 100 mph, past a screen of pine trees and into the sharp southern loop.

"For God's sake, slow down!" Salzer cried, almost thrown from his seat.

Rudi roared out of the curve, then back up the parallel straight. Only a thin patch of forlorn grass separated the two. He thought he had lost the race already but was determined to continue. In the 180-degree steeply banked northern loop that formed the eye of the circuit's needle shape, Rudi fought to make up the gap.

By the third lap, he had closed to the middle of the pack. By the next, a drizzle had turned into sweeping torrents that soaked him to the bone and cut his visibility almost to nothing. He barely slowed over the asphalt now slicked with rain and oil. His tires shot out plumes of water.

More and more cars withdrew from the race.

Coming into the northern bend in his seventh lap, Rudi spied the other white Mercedes on its side, the timekeepers' booth smashed into toothpicks. Rosenberger had skidded out of control: two race officials were dead, a third grievously wounded. The Mercedes driver and his mechanic were hurried off in an ambulance.

When Rudi stopped in the pit for gas and oil, he asked Porsche if his teammate was okay. "Just slightly hurt," Porsche lied.

Rattled nonetheless, Rudi slung out of the pit, back into the race. One lap after the next followed. He lost track which number he was on. Rain poured from the sky. On the opposite straight, French champion Jean Chassagne lost control of his Talbot, shot across the separating grass in front of Rudi, then rammed into the spectators.

Charly was in the grandstands, near the French driver's wife, when the loudspeaker announced that it was her husband who had crashed. Her face blanched and she wept quietly. Overwhelmed by the scene, Charly left.

Rudi advanced through the storm. He passed car after car but had no idea of his position. At that time, there were no signals from the pit for drivers as they passed. Even when there was a pause in the rain, he did not know where he stood. He drove through a haze of tension and exhaustion, Salzer urging him "Faster!" Onward he went down the straights, deftly maneuvering through the slick loops in between.

Finally, after twenty laps, 243 miles, and almost three hours of driving, he came into the finish, stopped, and lifted off his goggles. He was clueless as to where he had placed. His legs shook in the cockpit, almost like they were lost without the need to shift, brake, or accelerate. People crowded around him, heralding "Rudi — victory!" Even as they draped flowers over his steaming hood, he could not believe it. He had won his first Grand Prix and with it the sobriquet "The Rainmaster."

Later, at their hotel, Charly asked if he felt he must continue to race. There was too much danger to it all. "Yes," he said. "I must."

Four years after he earned the name "The Rainmaster," Rudi was sitting at the breakfast table in his mountainside chalet in Arosa, Switzerland. He stared at the letter that had just been delivered. It was stamped with the star and laurel symbol of Mercedes-Benz and dated November 1930. Each fall, the most prominent factory teams recruited their drivers for the following year. Offers were made, and contracts negotiated. Since his first AVUS win, Rudi had become one of the top aces in Europe, evidenced by the wall in his chalet lined with trophies and silver plaques. He was famous, rich, and confident of receiving terms that would allow him to live idyllically during the wintertime in Switzerland.

Now, though, he found himself without a team. The letter was from Wilhelm Kissel, Daimler-Benz's CEO, and it stated that his contract would not be renewed because of the economic crisis. Mercedes was abandoning racing. Put simply, the sport was too expensive.

The Wall Street crash of the year before had sent ripples of destruction — economic and political — far beyond the shores of the United States. Having barely recovered from its debilitating postwar hyperinflation, Germany was particularly hard hit. Banks were shuttered, and financial panic stalled production. Exports dried up. Corner shops, farms, industrial firms, local governments — all went bankrupt. Suicides spiked. Soup kitchens opened on street corners, but even these ran out of food. The desperate bred rabbits or planted meager home gardens.

Cushioned by his racing winnings and his own family fortune, Rudi

MERCEDES BENZ CLASSIC ARCHIVES

Rudi and his wife, Charly, in their heyday, traveling the circuit

had felt little effect from the crisis until now. Accompanied by Charly —now his wife—he traveled to Stuttgart to buttonhole Daimler-Benz management and attempt to change their minds.

He met with Kissel alone in his office. The forty-five-year-old executive was tall, slender, and sported a well-trimmed mustache. Before the 1926 Daimler merger, he had worked his whole career at Benz & Company. After taking over the helm of the combined company, he had led it to new heights with his skilled eye for numbers, talent, and sailing the political winds—until the worldwide depression hit.

No matter what argument Rudi made, Kissel ticked off the reasons why racing was a luxury he could ill afford: a plummet in domestic sales; the dead export market; a labor force reduced by half; a sagging bottom line.

Defeated, Rudi left to see Alfred Neubauer, who was waiting with Charly in his office, a room lined with overstuffed file cabinets. Although only eight years older than Rudi, Neubauer had become something of a surrogate father. Likened to a "modern-day Falstaff," he was big, round, and fleshy-faced. He loved to drink, eat, charm, tease, and put on a show. He could also be vain. But there the similarities with the Shakespearean character ended.

In the mid-1920s, after a hill-climb race in Austria in which his teammates, including the newbie Rudi Caracciola, had shown him their dust, Neubauer realized he might be better served in another job. His fiancée, Hansi, was blunter: "There was only one thing wrong, Fredl. The others drove like mad, but you drove like . . . a night watchman."

Neubauer had always been a better tactician than driver. He was also a master organizer with an elephantine memory for courses, drivers, and lap averages, and he knew every car inside out. After his embarrassing hill-climb loss, he traded his race goggles and overalls for a trench coat and a necklace of stopwatches to become the team manager. He reinvented the role, not least with his system of flag signals and chalkboards that alerted a driver to his position against his competitors, the number of laps remaining, and when to come in for a pit stop. Revolutionary on the Grand Prix circuit, his system was inspired by the 1926 German Grand Prix that Rudi had won without even realizing it. After that race, the two became almost inseparable, and by working together they had put together a string of victories almost without equal.

Now Rudi slumped in Neubauer's office. "It's finished," Rudi said. "All over."

Charly rose from her chair. "If you can't drive for Mercedes, then drive without them." Neubauer scowled dubiously. Charly suggested Alfa Romeo. Neubauer thought it a ridiculous idea. Rudi was solemn. No matter what, he needed to race. It was everything. Neubauer grew frantic. His cheeks mottled. "But . . . you can't . . . It's like going over to the enemy. It's almost treason."

"You can't expect me to give up my career as a racing driver."

After Rudi and Charly departed, Neubauer, a force of nature on an average day, worked on Kissel. If running a team was too expensive, what if Rudi represented Mercedes alone in the 1931 season? If investing in the development of a new car was too costly, allow Rudi to buy the latest SSK himself—at a discount, of course. Neubauer volunteered to manage Rudi and secure a small staff. If the company funded their salaries and expenses, they would split any earnings with Rudi. Mercedes could not lose—and given that Rudi had won almost a dozen events in 1930 alone, they might come out ahead. He would quickly prove the wisdom of the company's limited investment.

On April 12, 1931, the northern Italian town of Brescia was alive with the revving of homegrown engines: Itala, Bianchi, Maserati, Fiat, Alfa Romeo. The white Mercedes SSKL owned by Rudi was one of the few challengers of foreign design. Then ninety-eight cars shot down the road toward Cremona, the first stretch of the Mille Miglia, a 1,000-mile contest of amateurs and professionals that harked back to the intercity races from the turn of the century.

Following a figure-of-eight course over ordinary roads, with Bologna at its center point, the competitors swooped down the length of Italy as far south as Rome, then climbed north along the serrated spine of the Apennines back to Brescia. All along the route, from the biggest cities to the smallest hamlets, spectators lined the course to watch. Italy was motorsport-crazy, and the Mille Miglia was its national pride.

The route, raced night and day, was too long to reconnoiter in advance. It crossed through scores of villages, barren valleys, and nameless intersections. Only Italians, its organizers believed—and history had proven them right—could know more than a fraction of the often treacherous roads or muster the support needed for such a long race. They clearly had not seen the likes of Alfred Neubauer. He had charted out every mile and set up rolling depots along the course for the only driver under his management that season.

In his Mercedes, Rudi had the power advantage, and he barreled down the straights at over 120 mph, leaving the Alfa Romeos piloted by Italians Tazio Nuvolari and Giuseppe Campari in his rearview mirror. Rudi sustained this early lead through the first 200 miles. South of Siena, darkness fell, and somewhere along the serpentine roads to Rome, Nuvolari overtook him and claimed first position. Winner the previous year, Nuvolari was known by many nicknames, including "The Maestro" and "The Flying Mantuan." Such was his desire to dominate he had once competed in a motorcycle race with two broken legs bound in plaster—and won.

Over the course of the night, Rudi fell to sixth place, then farther back. Refusing to give up, he rallied and returned up the leaderboard.

Near Ancona on the Adriatic Sea, past the halfway point, he regained third position. But it looked like Alfa Romeo had the race locked. The fiery Nuvolari was in first. Campari, a jolly, pink-skinned bear of a man known for his eating binges and operatic voice, was second. Twice he had won the Mille Miglia.

Dawn crested on the horizon, and through a clinging mist Rudi edged his SSKL faster. His Italian rivals tried to box him out on an open stretch of highway, but he overcame them. With 250 miles left, he did not look back, crushing his nearest challenger Campari to the finish by over eleven minutes. It was the first win by a German driver—and a German car—in the Mille Miglia.

Shortly after Rudi's return from Italy, Kissel asked him for a favor. They were late on the delivery of a custom Mercedes because of its complicated build. To smooth over any potential aggrievements with their very special Munich-based client, Kissel wanted his celebrity driver to deliver the car personally.

Rudi drove the 7.7-liter, black convertible Mercedes with bulletproof windows, steel-plated side panels, and a glove compartment with a secret nook for a revolver from Stuttgart to Munich, stopping at the local Mercedes-Benz dealership, which was run by Jakob Werlin. There, the car was washed and polished to a shine.

Accompanied by Werlin, Rudi brought the Mercedes to a stop at the appointed hour of 5:00 p.m. in front of a grand stone mansion over which a swastika flag flew. Rudi was wearing a nice suit and tie, the better to make an impression on the client: Adolf Hitler.

Before the September 1930 Reichstag elections, the Nazi party leader had barnstormed across the country, promising to make Germany great again. Given a democratic government riven with dissension and the crippled economy, he found a welcome audience for his rallies. As one of his earliest supporters openly boasted, "All that serves to precipitate the catastrophe . . . is good, very good for us and our German revolution." When officials tallied the election results, the Nazis had emerged as a stunning new force. Before the vote, they held a meager 12 seats in the Reichstag, making them the smallest party. Afterwards, they commanded 107 seats and were the second largest.

Although Hitler's incendiary ideas and methods were well known, Rudi was eager to meet this rising power in Germany, if only as a curiosity. Reichstag elections were not something that much concerned the race car driver.

Awaiting them on the steps of the Brown House, the Nazi Party's national headquarters, was Hitler's private secretary, Rudolf Hess. Hess led Rudi and Werlin into an expansive, high-ceilinged study where Hitler sat at a desk. Behind him on the wall was a life-sized portrait of Henry Ford. Hitler was a car enthusiast, and he greatly admired the American tycoon.

Hitler rose immediately out of his chair. Rudi had seen numerous photographs of him, but the man was shorter and stockier than he had imagined he would be. Hitler lavished praise on Rudi for the Mille Miglia win. What a victory for Germany! Before Rudi could respond, Hitler rattled a barrage of questions at him about Italy: How were its trains? What were the living conditions? Did people admire or loathe Mussolini? Rudi made a joke about how little one could see at 100 mph, but Hitler pressed.

He seemed uninterested in the brand-new Mercedes parked outside and instead gave the two men a tour of the Brown House. He showed them the bloodstained banner from the Beer Hall Putsch in 1923, where the Nazis made their violent bid for power, and the Senators' Hall, a roomy space with red-leather chairs and an elevated platform where Hitler spoke to his disciples. Finally, they saw the basement and its corridor of massive steel safes that contained the Nazi Party's files, including member cards. "Reached 500,000 today," one of his staff remarked.

Growing impatient, Rudi got around to the Mercedes. "I've come to demonstrate it for you. It turned out to be a beautiful car." Instead of going with them, Hitler promised to come to the Mercedes garage in half an hour, and Rudi and Werlin left. On schedule, Hitler arrived with several guards and a man introduced as Julius Schreck. Musclebound, with the slitted eyes of a snake, the ex-soldier was the first leader of the Schutzstaffel (SS), a Nazi paramilitary force. Werlin pointed out the characteristics of the Mercedes while Schreck advised Rudi not to drive faster than 20 mph: "The enemies of our party would promptly use an accident for counterpropaganda."

With this warning in mind, Rudi chauffeured Hitler around Munich at a quiet pace. Marveling at the car, Hitler stated that it was a true testament to German craftsmanship. Rudi returned the Nazi leader back to the dealership to finalize its delivery.

On leaving Munich, Rudi reflected that there was not much in Hitler's presence — or looks — that destined him for momentous things, but the meeting still made a deep impression on him. Whatever the future held, the Nazi leader was a sure supporter of racing, and this alone mattered to Rudi.

Although he had won the German Grand Prix and several other races across Europe after the 1931 Mille Miglia, he and Mercedes parted ways. The worldwide depression continued unabated, and the plummet of auto sales made any sponsorship, even a profit-sharing one, difficult to maintain. Again, Neubauer tried to keep Rudi in the star-and-laurel family, making promises about competing in the United States. Without any guarantees, and knowing Mercedes would not invest in keeping up with other automakers, Rudi declined. "I've got to drive, don't you see," he told Neubauer before signing with Alfa Romeo.

Winning was the only thing on his mind when he joined the Milanese factory team for 1932. After spending so much time in the 3,600-pound Mercedes SSKL, a car he likened to manhandling a locomotive, he could barely contain his joy the first time he drove an Alfa. "As light-footed as a ballerina," he praised its handling to its designer, Vittorio Jano. His new teammates, including Tazio Nuvolari, were skeptical that the German could handle such an agile, featherweight car, but Rudi quickly proved them wrong. The 1932 Eifel win over René Dreyfus was his first noted triumph for the Italian make. Others quickly followed.

In their Alfa Romeo P3 *monopostos* (single-seaters), crimson, supercharged bullets on wheels, Rudi and the Flying Mantuan dominated the Grand Prix. Tazio won the French. Rudi claimed the German. He also triumphed at Monza and a host of smaller races and hill climbs. For the season, Rudi ranked second only to his teammate, who won the European Championship. They were truly on top of the world.

3

The Speed Queen and Old Gaulish Warrior

JACQUES MARSILLAC, A reporter for the popular Parisian daily *Le Journal,* arrived at a stylish Parisian restaurant ready for lunch. A distinguished correspondent who had covered insurrections in Ireland and North Africa, Marsillac, for his next assignment, would accompany American heiress and rally driver Lucy O'Reilly Schell on latest next adventure. Trench coat draped over his arm, he surveyed the lunchtime crowd.

A woman's voice rang out over the hubbub. "So what if their car went into the ditch four times?" it boomed, in English. "Was that really a good reason to forfeit?" Thirty-five years old, thin, ruddy-faced, blue-eyed, with bobbed auburn hair, Lucy Schell commanded the room. In fact, she commanded any room she occupied, all five feet, four inches of her, whether she was dressed in haute couture in a city restaurant or in oil-stained overalls in a garage. Her husband, Laury, to whom she had addressed her remark, did not protest. Laury was Lucy's opposite in every way—as reserved and quiet as she was lively and locquacious.

Lucy spotted the *Le Journal* reporter across the restaurant. Her bright smile widened as she approached him, and in flawless French, she said, "At last! There you are. Come, I'll introduce you. We were talking about the Rally." Once seated at the table, Marsillac got an earful about what lay ahead in his role as "ballast" for the Schells' car in the upcoming 1932 Monte Carlo Rally.

Launched in 1911, the Rally was a supreme test of endurance for its participants and their automobiles. As one journalist wrote, "There comes a moment when respectable drivers of unblemished reputation see imaginary elephants in pink pajamas wandering on French main roads where no elephants should be. That is the crux of the run." Each winter run brought a new wrinkle in how to push both cars and drivers to their limits. For the 1932 Rally, competitors would start from their choice of nineteen far-flung places, including Stavanger, Norway; Gibraltar off southern Spain; Athens, Greece; John o'Groats in the north of Scotland; and Palermo, Sicily. As always, the finish line was in Monte Carlo.

The greater the distance traveled, the greater the challenge and the higher the number of points to be earned by the drivers. Drivers needed to reach control stations during a specific window of time to prove that they were maintaining an average speed of at least 25 mph. At first blush, this seemed slow to Marsillac, but Lucy explained how difficult it was to meet even that pace given that no allowances would be made for sleep, food, refueling, repairs, poor navigation, or accidents.

The Schells planned to tackle the second-longest route, starting in Umeå, Sweden, 100 miles from the Arctic Circle and some 2,300 miles from Monte Carlo. The almost nonstop journey would take four days and three nights. When the teams arrived in Monaco, they would have to undertake a convoluted series of tests of their driving skills — and the reliability of their cars — to determine the ultimate winner. Although he was no delicate flower, Marsillac was cowed by such an adventure, particularly when he learned about the icy road conditions. That the pilot was a woman made him even more uneasy. Fortunately for him, he kept that fear to himself — nor did he ask why a mother of two adolescent boys was taking part in such a treacherous competition. Had he done so, he would have had to find another ride.

A week after the lunch, at five in the morning, the Schells arrived at the Le Journal's offices to collect Marsillac. They appeared ready for a polar expedition. Lucy wore a long waterproof jacket, wool trousers, and tawny leather boots that reached her knees. When they had met for lunch the week before, Laury had looked to Marsillac like a sallow mortician. Now, in his heavy, fur-lined coat and boots, he seemed a vigorous giant.

Their black Bugatti T44, a midsized tourer with filleted lines of silver and red down its body, was similarly well fitted out, with mud fenders, spare tires lashed to its sides, and three headlights perched on a bar at the front. Inside, there was enough gear to mount a siege: food stores, mechanic's tools, picks, shovels, rope, tire chains, and a block-and-tackle set that could lift the car out of a ditch. There was so much stuff that Marsillac had to burrow himself a space on the back seat underneath a pile of luggage, blankets, and camera equipment. He was joined there by Hector Petit, winner of the 1931 Rally, who was catching a ride to the starting line in Umeå. Petit joked, "Never Nanook, leaving for the North, had his shoulders laden with more."

"When do we sleep?" asked Marsillac.

"When everyone is too tired to drive," Lucy answered from behind the wheel. At those times, they slept inside the Bugatti on the side of the road.

Lucy drove as much as her husband. She was every bit as resilient — and a bit faster too. After five days and nights, they were on the last leg of their journey to Umeå. A number of competitors had already turned back, giving up even before the start of the race. One rallyer never had even that chance. While swerving to avoid an approaching horse-drawn sledge, he overturned his car and was killed.

The Schells and their passengers had just left Sundsvall, a couple of hundred miles from Umeå. Lucy dozed in the back seat, Laury was driving, and Marsillac was navigator. He dared not risk piloting the car himself. It was still dark, and a thick fog had settled over the road, which was covered with four-inch-thick black ice. "Are you sure we're on the right road?" Laury asked, yet again. "Shouldn't we be turning?"

Marsillac stared at the map, which was illuminated by the bluish glow of the onboard lamp. "No, we're on the right road." Seconds later, their tire chains lost their grip, and the car swung to the right. There was a brief moment when it seemed like they were gliding through air. Then they smacked into a snowbank.

When everyone inside had regained their wits, they realized that the car was tilted on its side. Any movement made it rock. Carefully the four crawled out and inspected the damage with flashlights. The car looked okay, but it was half submerged in the snow and their winch was insufficient to the task. It was 1:00 a.m., 30 below zero, and they were stuck in the middle of a pine forest at least a dozen miles from the nearest village.

Petit and Marsillac lumbered off in a biting wind to try to find a farmhouse and ask for help. Lucy and Laury hacked at the hardened snow with their picks and shovels. One hour passed. Then two. Petit and Marsillac returned, having given up the search after they heard wolves in the woods. The Schells had barely made a dent in the snowbank, yet kept digging. Any thought or comment about abandoning the race before the start, as others had done, was never even raised. This was just an ordinary day for the Schells.

Miraculously, two forest workers appeared down the remote road and promised to return with help. An hour later, a truck fitted with a huge snowplow arrived. Then another truck came. Ten Swedes with rough, bearded faces and fur caps cleared away the snow underneath the Bugatti and hauled it free from the bank. The stranded rallyers garbled thanks in pidgin Swedish, then started the car. The engine puttered uneasily in the bracing cold, but then it settled.

They advanced again into the night, getting lost a few times over the last 100 miles to Umeå. Roads splitting off in small towns often went unmarked. One local barricaded his door when the Schells knocked to ask for directions, but most people generously pointed out the way.

Lucy was at the wheel for the final stretch, driving through what looked like a tunnel of snow as the headlights cut through the blizzard. This was not the life her parents had wanted for her, but saying "yes" to Lucy was always a far cry easier than saying "no."

The only child of wealthy parents, Lucy had received every advantage, to the point of being spoiled. What Lucy wanted, Lucy got, but this doting instilled in her a confidence to pursue her ambitions rather than settling her into indolence.

Her father, Francis Patrick O'Reilly, was raised in Reading, Pennsylvania, the son of Irish immigrants who fled the Great Famine of the 1840s. He made a fortune, first in construction, then by investing the proceeds in real estate and factories in his hometown. At forty-six, he was rich and ready for his life's next chapter: traveling the world. In January 1896, he married Henriette Celestine Roudet in her hometown of Brunoy, south of Paris. A quick nine months later, in a Parisian hospital, Lucy Marie Jeanne O'Reilly sprang into the world.

Lucy spent her youth traveling between the United States and France. She was decidedly *nouveau riche* and unapologetic about it. A biographer wrote of her blossoming personality: "While she grew up in the United States and absorbed its spirit of independence, she remained unmistakably Irish both in looks and temperament, combining a natural charm and vivacity with headstrong courage, obstinate determination, and a careless outspokenness." When asked the country to which she swore the most allegiance, she said, "I am American," but the briskness of her answer betrayed the feeling that she never felt completely at home anywhere. In Paris at least, she felt the sting of the general attitude: "When one is not French, one is a foreigner."

Her Grand Tour—the upper-class ritual of exploring the art, culture, and polite society across Europe—came to an untimely halt with the start of World War I in 1914, but her jaunt had not been without consequence. While traveling , she met Selim Lawrence "Laury" Schell.

Although his parents were American, Laury was French in spirit. A diplomat's son, he was born in Geneva, Switzerland, and his family had settled a short distance from Brunoy. Laury trained as an engineer but was uninterested in work, despite having only a meager inheritance. Lucy's father tried to convince her not to get serious with him, saying, "His life seems to consist entirely of the pursuit of pleasure," but Lucy was not one to listen to advice.

In the early part of the war, she labored enough for the two of them. Working as a nurse at a military hospital in Paris, she helped treat soldiers who had suffered every type of horror, particularly injuries from artillery shells: severed limbs, burned flesh, disfigured faces, and bodies riddled with shrapnel. The sight of their wounds—and their suffering—was seared into her mind.

In April 1915, a month after Zeppelin ships began bombing Paris, Lucy and her mother, accompanied by Laury and his brother, left for the United States. When interviewed by a Reading, Pennsylvania, newspaper, Lucy railed against the calamity brought by the German invasion and thanked America for its aid. She also promised that France was not yet defeated. "Even the most dangerously wounded soldiers as they lay on their beds of pain and tossed and moaned in delirium begged and prayed to be allowed to return to the front and fight for their beloved country."

Two years later, Lucy and Laury came back to France and wed. After the Armistice, they lived in Paris, a place and a time that Ernest

Hemingway famously described as "a moveable feast." One of the flood of American expats who inhabited the city said, "Every day was like a sparkling holiday." Bohemian parties in Montparnasse. Booze and art, literature and sex, cafés and cabarets. Paris bloomed with the talents of F. Scott Fitzgerald, Cole Porter, Ezra Pound, Coco Chanel, Man Ray, and Pablo Picasso. Only Berlin challenged the city in its effervescence. It was a grand moment, and Lucy was one of the cadre of rich Americans fueling the party. The births of her children — Harry in 1921, and Philippe five years later — failed to settle her down. Instead, becoming a mother had the effect of revving her up.

A defining characteristic of the Roaring Twenties was a love of speed. Across Europe and the United States, car races of every sort proliferated: "flying kilometer" trials, hill climbs, circuit contests, and intercity rallies. Laury and Lucy were both drawn to the scene, first as spectators, then as drivers. Thanks to Lucy's family's money, she could afford to buy the latest, and best, cars. And she could compete in those cars as well — if and when she was allowed.

Lucy followed in the footsteps of other groundbreaking speed queens like "The Godasse de Plomb," Violette Morris, and "The Bugatti Queen," Hellé Nice. They proved themselves to be particularly adept at endurance races and long-distance speed records, whether they were racing against the men or only other women.

In 1927, Lucy signed up for her first race at the inaugural Journée Féminine de L'Automobile at the Montlhéry autodrome outside Paris. "Ladies Day" was a big hit, and Lucy caught the bug. By the early 1930s, she was one of the top female drivers in Europe, having racked up a number of wins in the Bugatti T35, Talbot M67, and Alfa Romeo 6C. She was seldom photographed in her race overalls; instead, newspapers and magazines featured her dressed to the nines, wearing high heels, mink scarves, and pearls. The press wanted her to drive fast and to be ready for the runway an hour later. Lucy was game for the show. As for silver-spoon airs or a lack of toughness, she suffered from neither. Before one race, she broke her arm in several places, and her doctor advised her to forfeit. Lucy refused, participating with a thick plaster cast on her arm and nearly winning.

Her favorite event was the Monte Carlo Rally, a challenge that, as one chronicler remarked, appealed to those "looking for trouble." In 1929, she ran it alone and placed eighth overall and first among the women, winning the Coupe des Dames. The next two years she partnered with Laury. In 1931, departing from Stavanger, they placed third. The two always flew the Stars and Stripes, and year after year they were the rally's highest-placed Americans.

The *pop-pop* of flashbulbs lit up the early hours of Saturday, January 16, 1932, as photographers captured a smiling Louis Chiron before he sped away in his Bugatti from Umeå, at 3:34 a.m. At intervals, one brand of car after another followed him over the starting line: Riley, Sunbeam, Lagonda, Triumph, Ford, Studebaker, Chrysler, and many others. They were cheered on by a few hundred Swedes, most of whom had traveled there on skis or ice skates. A race official announced to number 57, the Schells' Bugatti, "Two minutes, gentlemen."

Lucy ignored the "gentlemen." She waited at the wheel for the signal to go. Once they were off, she drove steadily and slowly, very differently from the others, who had all torn away at full throttle. The road was coated with a foot of ice, its surface glistening. One could barely stand, let alone drive, on it without spikes or military-grade tires. An indelicate turn of the wheel would guarantee a pirouette into the ditch. Indeed, Marsillac noted, to crash would have been as easy as to "drink a Vermouth or a cup of tea."

A Talbot swooped past them. Before it disappeared around a bend, its headlights danced left, then right. There was a sharp swish, almost like the sound of a boat breaking through a swell, and the lights were extinguished. Moments later, Lucy pulled up alongside the car, which was stuck deep in a hollow beside the road. Its team waved at them — all okay — and Lucy drove on. Save for injury, regulations forbade drivers to help other competitors.

Despite several sideway lurches of their own into ditches, the Schells' car reached Sundsvall fifteen minutes before the cutoff time. At the control, located inside the town's fanciest hotel, they dashed off their signatures, devoured some ham sandwiches, and started toward Stockholm, 230 miles away. A mix of rain, snow, sleet, and fog met them — and this on narrow roads through ravine-ridden countryside. From the back seat, Marsillac likened the jolting, back-and-forth movement across the road to being stuck inside a cocktail shaker. Even his brain hurt.

They shared the road with local traffic: motorbuses, sledges, and sometimes families out on ice skates, which made the journey all the more maddening. Use of the brakes guaranteed an uncontrollable slide, but both Schells handled the Bugatti with consummate skill. They rarely slowed, allowing the wheels to glide across the ice to carry them around turns. Marsillac was particularly struck by how Lucy's smile widened the tougher conditions became.

At 1:50 a.m. on Sunday, they arrived in Stockholm and took a brief break to stretch their legs and refuel at the depot. Over a dozen teams had already retired from the race. Then they were off again, cutting southwest across Sweden's pine forests on their way to catch a ferry to Denmark. They were exhausted but drove onward through a blinding snowstorm. Every so often they passed a car stuck hood-deep in a ditch, its team trying to wrest themselves free.

Hour after hour of navigating these roads, only able to see a few yards ahead, wore down the drivers. Nerves deadened. Concentration waned. Errors multiplied. The carnage beside the road was evidence enough of that. The Schells' Bugatti spun out of control three times in as many hours; fortunately the only consequences were some dings to their coachwork and thrown tire chains.

At 6:00 a.m., a few hours before the Nordic sunrise, they decided to rest for a short spell. They parked the car by a barn and dozed off quickly, leaning against each other in the front seat. Marsillac squirmed about until he found a comfortable position in the back. Then—

"It's seven! Quick, back to it," Lucy announced, clapping her hands.

Soon after, their headlight mount sheared off. They could not continue in the dark. Summoning help was no simple matter: even if they could find the nearest town in the dark, it was Sunday, everything was closed for the weekend, and none of them spoke Swedish. After some cursing, the Schells finally jerry-rigged the mount back in place with some wire and hope. The repairs took time—too much time. By their maps, they had over 150 miles to go to the ferry port at Helsingborg, where they needed to arrive before the last ferry departed at 1:45 p.m.

"We are not stopping again," Lucy declared. It was already past 9:00 a.m. Hell-bent, she whizzed at 40 mph along the axle-deep rutted roads that would have best been navigated at a crawl. Fortunately, they now had the light of day by which to travel. The miles passed in a blur, as did the massive forest alongside them. As they neared the coast, Laury took over the wheel while Lucy continually checked her watch. One hour until the ferry left.

Forty-five minutes.

Thirty minutes.

Fifteen.

Ten.

She sighted Øresund Strait, then the ferry.

Laury whipped down the road. Their surviving competitors were already loaded onboard. Plumes of black smoke pumped from the ferry's stacks. Marsillac checked his own watch. He felt like Phileas Fogg risking everything to travel around the world in eighty days, only to almost fail because of a measly ferry schedule.

The Bugatti came to a jarring halt by the gangplank, and Lucy leaped from her seat, shouting, "Wait, Captain! Stop!"

They had made it.

Onboard, despite the shuddering din of the ship's bow breaking through gray ice floes, they curled up and slept for the half-hour voyage.

At Odense, Denmark, they gleaned news from the drivers who had started from Stavanger. They too had suffered terrible roads, and only two teams had made it from Norway. One driver's car was being held together with an elaborate bowtie of string and copper wire. Gaunt and hollow-eyed, every last competitor looked like they had been touched by the hand of death.

The Danish roads proved better, and the Schells and Marsillac reached Germany without any issues. At the frontier, they guzzled black coffee before speeding off toward Hamburg. From there, they turned toward Brussels, 375 miles away. Such was their exhaustion that it was a challenge to recall what country they were dashing through at any moment. Over a week had passed since any of them had slept in a proper bed, bathed, or even removed their boots. They were filthy, and their clothes reeked. When Marsillac went into a hotel to send a telephone dispatch to *Le Journal,* the staff looked at him as if he were a beggar. Then it was back to the car.

Hour after hour, day after night, Lucy and Laury drove. Naps on the back seat barely dented their weariness. At this stage of the rally, the term "endurance trial" had taken on new meaning. Even when Lucy was not piloting, she took care to make sure her husband remained awake at the wheel. She pestered him with questions whenever she feared he might be drifting away.

"Laury, don't you think it's a little icy?" she asked on the road to Belgium.

"No," he answered. "That's melted snow. It's not slippery." She looked at him. "I swear," he said softly. "But if you really want, we'll take it slow."

From Brussels, they journeyed 200 miles to the Paris checkpoint, arriving in the afternoon under dark clouds and a persistent drizzle. They needed to reach Monte Carlo the next day, between 10:00 a.m. and 4:00 p.m. Over 600 more miles to go.

The highway south of the French capital was a slippery mess. Lucy and Laury reduced their shifts to an hour each time, yet their reactions at the wheel remained mushy. Their speech slurred; their eyes drooped asleep for tenths of a second before being wrested awake. No amount of fresh wind from an open window, no amount of shaking their heads, no amount of clenching their fists, could alleviate the weariness.

They continued down the poplar-lined road, often through a heavy downpour. Whenever Laury asked Lucy to rest, she would grumble that she was fine but would eventually give over the wheel. There were little mistakes: a bend taken too wide, a drift to the left, a missed turn. These piled up, but none of them resulted in an accident.

Others were less fortunate. A Danish team, while they were fixing a broken headlight at the roadside, was struck and killed by another competitor blinded by the rain and his fogged-up windshield.

On Wednesday morning, they crossed over the mountains down into Monaco. Rain pounded the pavement, and fog obscured the beautiful terraces of Monte Carlo. When the Schells emerged from their Bugatti at the finish line, on schedule, they swaggered like conquering heroes. Only half the contingent from Sweden had made it.

After a short rest came the "flexibility test" to decide the winner. The team who traveled 100 meters in the slowest time with their engine in top gear followed by 100 meters in the quickest time would earn the most ad-

Lucy Schell and her husband, arriving in Monaco at the end of the
Monte Carlo Rally, 1936

ditional points. After many hard-fought miles, it was a farcical way to
determine the victor, but that was the Rally: never quite fair, and always
fraught until the end.

Laury piloted the Bugatti during the test, Lucy by his side. She was
so tense that she forgot to start her stopwatch. As the Bugatti puttered
at 3 kph in fourth gear across the 100-meter stretch, she urged her husband to give the engine some gas, believing surely that it would stall.
But Laury managed to crawl forward without incident. Then he turned
in a semicircle, jammed on the gas, and sped back across the same distance. Maurice Vasselle, a fellow starter from Umeå, scored with the
best times in his Hotchkiss AM2 and claimed first place. The Schells finished seventh.

Caught up in the spirit of the Rally, and awed by Lucy Schell, Marsillac chronicled their adventures for all of Paris to read. He concluded
his five-part series: "The dream has ended. Life must resume." His story
shared the front page with news of a Paris train derailment, another

shake-up of the cabinet, and a story about the "nationalist leader," Hitler, who was promising that "only by his own strength" would Germany rise again.

Secured from these troubles by her fortune—for now—Lucy thought only of finding a car equal to her ambitions to be the first woman to win the Monte Carlo Rally.

Her timing was impeccable.

The tall, distinguished man in a classic gray suit and round felt hat walked through the Left Bank. It was the spring of 1932, and the gardens about Paris were abloom with cherry trees and beds of tulips. Charles Weiffenbach, however, was not one to smell the flowers. Striding straight as an iron rod, the lantern-jawed sixty-one-year-old looked like a general who had misplaced his uniform.

At 59 Boulevard des Invalides, "Monsieur Charles," the longtime production chief of Delahaye Automobiles, pushed through the front door of the apartment building. He had a lot on his mind as he prepared to meet with the company board. There were big decisions to be made about its future, if it had one.

"The thirties were doom-laden," wrote one historian. "Nothing is more evocative of France at this moment than the picture of the grand old lawyer Poincaré, the national leader who had done so much to pull his country through the First World War and its aftermath. Worn out and ailing, he retired to his country house in Lorraine, and, opening the window with trembling hands, would look anxiously at the eastern horizon. He felt 'they would come again.'"

Weiffenbach may not have feared the Germans that morning, but there was much to trouble him in the decade later known as "the hollow years." Bumbling politicians were struggling to contain a tide of problems: spiking taxes, an unbalanced national budget, rising unemployment, xenophobia, and poisonous social problems.

Only a few years before, the French auto industry had boasted over 350 manufacturers; now only a few dozen remained. The Great Depression and the subsequent collapse of the export market had decimated their ranks, and even the Big Three—Peugeot, Renault, and Citroën— were on the ropes. The stolid, reliably conservative Delahaye company, known for building vehicles of consistent quality, was in dire straits, its production of cars and trucks down to a fraction of their former number. Something needed to be done or Weiffenbach would have to institute broad job cuts.

Madame Marguerite Desmarais welcomed Weiffenbach into her elegant apartment. Her brother, Georges Morane, who had partnered with her late husband Léon to buy out the company's eponymous founder in 1897 was there too. Georges and Marguerite were now the lead shareholders, but they had long ago given over day-to-day control to their nephews, who were also board members and who were present at the meeting.

Gathered at the dining-room table in the apartment, the board put two options to the vote. First, Delahaye could abandon cars altogether and focus instead on its line of utility vehicles: trucks, buses, and fire engines. Second, they could pursue an aggressive policy of expansion, hoping that the cost reductions from mass production would carry them through to brighter days. Prior to the meeting, Weiffenbach and the board had investigated both options, and it was in the nature of the energetic production chief to push for expansion, capital-intensive though it would be.

After much deliberation, Marguerite Desmarais issued her opinion: "Midrange touring cars built in high numbers — you say you can't build them without major investment. Your uncle Georges and I can't invest on this scale, but we don't want to lose control of the company. So that's the problem, boys." Neither would only producing utility vehicles prevent the company from singing its swan song.

She proposed an inspired third option: "If you can't build automobiles in quantity, build fewer but better ones. Win races to make the marque better known and to sell more luxurious and expensive cars. That is the decision which we take today."

Marching orders in hand, Weiffenbach headed back to the factory at 10 rue du Banquier. Delahaye was returning to competition, where the company first made its name when founder Émile Delahaye participated in the 1896 Paris–Marseille–Paris race. His automobile company was only two years old at the time. Unlike other early French manufacturers, such as Peugeot, which used Daimler engines, Delahaye built his own from scratch, making his patented Type 1 model the first 100 percent French-made car when presented in Paris. The two-cylinder engine, horizontally mounted in the rear, was rated at 6 horsepower. The car featured electric ignition (versus the typical Bunsen burners heating platinum tubes), an automatic carburetor, and one of the first uses of a radiator to cool the engine. At its debut, the design won unabashed praise, but few sales.

Delahaye, who was fifty-three years of age and had previously built machines for the brick trade in Tours, decided that he needed to show off his Type 1 in a race. Moreover, he intended to pilot one of his company's two vehicles himself, weak constitution be damned.

The 1896 race started from Versailles on September 24 and covered a distance of 1,062 miles over ten days. On the first day, the thirty-two entrants headed down the macadam roads, sometimes in excess of what was then a hair-raising 20 mph. The next day thunderstorms and gale-force winds raged and trees fell across the roads along the route. A Bollée tricycle car crashed into one. Another blocked Émile Delahaye. He commandeered a saw from a local farm, separated the tree into three parts, and removed the middle section.

The delay dropped him to sixteenth place, but the other Delahaye, driven by aviation pioneer Ernest Archdeacon, remained in a solid sixth. The rains continued. Each day more cars dropped out, some because their engine's Bunsen burner couldn't maintain a flame, and others after a crash or, in one exceptional case, a head-on attack by an enraged bull.

Fewer than half the entrants completed the race. With an average speed of 14 mph, Archdeacon finished seventh; Émile Delahaye came in tenth. Most important, he had proved his car's durability, and a line of customers followed.

Late the following year, desirous of an expanded operation but lacking the physical health that would be required to achieve it, Delahaye sold his company to two Parisian industrialists: Georges Morane and Léon Desmarais. The Morane family ran a candle-making concern in Les Gobelins—a Paris district crammed with tanneries, dyeworks, and other industrial plants—and the new owners moved the automobile manufacturer from Tours into a factory there, on the rue du Banquier.

Before Émile Delahaye's retirement, he made one more notable mark on the company's future by hiring twenty-eight-year-old Charles Weiffenbach to lead production. What Weiffenbach lacked in experience he

compensated for with hard work and an ability to hire—and inspire—
the best. The company blossomed under his watch.

By 1902, the Delahaye company had dropped out of motor racing.
There was no need for the expense. The company produced a range of
new passenger cars that sold well in their own right. It had made a few
unconventional leaps ahead of most other firms—a shaft-driven transmission, an overhead camshaft, a V6 engine arrangement (two banks
of three cylinders at a 30-degree angle)—but Delahaye primarily built
very conservative vehicles known for their robustness. "Solid as a Delahaye" was the company's calling card.

Following his instinct to explore other markets, both domestically
and abroad, Weiffenbach also built trucks, taxicabs, postal vans, farm
equipment, and boats.

By 1913, the company was producing 1,500 vehicles a year, ranked in
the top ten of French automobile manufacturers, and provided healthy
profits for its owners. "Monsieur Charles" was decidedly the force behind its success. He was a no-nonsense leader, blunt as a hammer.
Weiffenbach kept his office close to the factory line and rarely stayed
behind his desk. He set the pace for his staff, literally, taking many
quick-stepped walks through the workshops. If he asked his workers
for overtime, he stayed late as well. They were like family to him. He
expected the highest standards in materials, precision tools, and production methods. He rooted out inefficiencies and employed economies of
scale at every opportunity. Few managers better exemplified the motto
"Tough but fair."

As the 1920s passed and inexpensive, mass-produced cars became the
standard, Delahaye continued to mint a range of pricey models defined
by what many generously labeled "sober elegance." In reality, the company was stuck in the past, as symbolized by its radiator cap: a helmeted
Gaulish warrior. Its automobiles were durable and safe, but they were
no longer remarkable.

"The typical product," wrote British motorcar historian Cyril Post-
humus, was a "somber, high-built sedan with claustrophobically small
side and rear windows, a rear trunk, heavy-hubbed wheels showing all
the nuts without even the grace of plated caps, cart springs, and a rice
pudding performance." French car historian François Jolly described
Delahaye's ideal customer as a bourgeois "provincial notary in a black

suit, white fake collar, black tie, and soft hat," who felt comfortable "seated solemnly in a square box without ostentation." In this period, Jolly concluded, Delahaye built "the perfect car to drive in a funeral procession." In other words, the company was headed down the road toward oblivion, accelerated by the worldwide slowdown.

Despite satirical comments from the press that someone had slipped "a Benzedrine tablet into Grandma Delahaye's *café au lait* and achieved an inspiring metamorphosis," Marguerite Desmarais had given Weiffenbach a sound plan to right the ship in early 1932. Her strategy echoed the one Ettore Bugatti had suggested to the Delahaye chief of production at about that same time. "Charles," he said, "for years you manufactured excellent machines, but they don't appeal to customers, and they have no speed because you make them as heavy as your fire department trucks. If your cars were lighter and faster, you could prosper again." Bugatti also advised Weiffenbach to race again.

Many years had passed since Delahaye had made any great design strides, let alone competed in motorsport. Few believed that Monsieur Charles would take such a risk, and fewer still believed that Delahaye would succeed in emerging, as one critic remarked, "from the shadows of a 30-year hibernation." Weiffenbach was willing to take the gamble if it meant saving the company.

The sun was shining brightly at Montlhéry, the famous autodrome outside Paris, when a peculiar-looking car bearing no indication of its make or model drove out onto the track. The chassis was long and stood high off the concrete. Its thin-skinned body was a flattened oval at the front and stretched out in the back like a beetle's tail. At the wheel, which was centered awkwardly forward, was Albert Perrot, a former test pilot and a race car driver of moderate success. Overseeing the last-minute tweaks to the engine was a nattily suited man of similar age with a bushy mustache and a gaze that always seemed to be trained on a spot far off in the distance. His name was Jean François.

Soon after the board meeting that decided Delahaye's change in direction, Monsieur Charles tapped François to lead the company into a new era of fast and agile cars. He was not given free rein to spend whatever he wanted, nor was he given unlimited time to study the problem. Typically, Weiffenbach expected François to produce a practical and economic solution in short order. The strange vehicle at Montlhéry that day in early 1933 was the first fruit of his efforts.

Under the hood was a 3.2-liter, straight-six-cylinder engine, the modified version of the Type 103 engine that Delahaye had first produced in 1928 for three-ton trucks, which their factories remained equipped to make. Tough, strong, and able to run for long periods at full throttle, the truck engine could be forced in a way few car engines could handle.

The "son of a truck" engine on the prototype car at Montlhéry was coupled with a light chassis featuring independent front suspension. This was the first Delahaye vehicle to allow the wheels to move independently of one another, an advance the company was already well behind in adopting. Its purpose was to keep the tail (wheels, tires, brakes, and assemblies) from unnecessarily wagging the dog (pretty much everything else on the car) so as to allow for a smoother, better-handling drive over a range of surfaces. Back when such a suspension was far from the fashion, François had designed one for his former employer.

Wearing a black cloth helmet and white driving suit, Perrot took off down the concrete track. The car quickly surpassed 70 mph, a speed within the capacity of most Delahayes at the time. Then the needle on the speedometer crept higher, and higher, and Perrot steadied out at almost 100 mph. For the next three hours, he maintained that average speed to complete an even 500 kilometers around Montlhéry's banked autodrome.

Other makers had fielded faster cars over similar distances, but for Delahaye the achievement was akin to magically turning a tortoise into a hare. Struck by the performance of the ungainly-looking car, a visitor at the track asked the attending mechanics, "What on earth was that?"

"Eh bien, M'sieur . . . That was our new Delahaye. It will be at the Paris *Salon*."

Held in the Grand Palais, just off the Champs-Élysées, the 1933 Paris Motor Show, or Salon de l'Automobile, was a smashing success for Delahaye. Weiffenbach introduced two sibling versions of their new design direction, both deemed "Super-Luxe," with sleek bodies and triangular front grilles created by a high-end French *carrossier*. Like many auto manufacturers, Delahaye produced everything but the bodies of their cars, leaving customers to order bespoke ones from the coach-builders of their choice.

The first version, Type 134, featured a four-cylinder, 2.1-liter engine and independent front suspension on a chassis with a short wheelbase of 112 inches. The second, more powerful version, Type 138, matched the specifications of the prototype tested at Montlhéry: a straight-six, 3.2-liter engine, with the same suspension as the 134 but with a wheelbase a foot longer. The French motorsport press celebrated the two designs.

During the Salon, Perrot even repeated his 500-kilometer performance at Montlhéry in front of a herd of reporters and spectators who had come out from Paris for the day.

It did not take long for buyers to appear. Lucy Schell was foremost among those who were captivated by the Delahaye presentation. She wasted little time showing up at the rue du Banquier factory.

Weiffenbach was working through some papers in his office when his secretary peeked in through the door to inform him that he had two visitors. They did not have an appointment but had insisted on seeing him nonetheless.

"Their names?" Weiffenbach asked.

"The Schells: Lucy and Laury," his secretary answered.

Before Weiffenbach could invite them inside, Lucy Schell burst into his office, her husband in tow. Weiffenbach knew them by reputation. Lucy was too big a personality in the Paris automotive scene to go unnoticed.

"You may have heard of us; I am Lucy O'Reilly Schell, and this is my husband Laury." She did not wait for a reply. "We like the look of your new Super-Luxe and want it for the Rally next season. The 138 is too big, so it will have to be a 134."

Beside her, Laury smiled sheepishly. He was accustomed to Lucy driving the conversation. She continued, explaining that the 134 would do for now as it was, but they thought that Delahaye should put the straight-six engine of the 138 into the shorter, lighter chassis of the 134 to improve its potential as a sports or rally car.

"Can I ask you," she said. "Are you considering that, at all?"

"Ah, *ça alors,*" Weiffenbach said, sensing that he was already in a negotiation. "I'm afraid that would be impossible. My engineers are far too busy to undertake a special project like that. In any case, I thought you drove a Talbot?"

"Talbot is finished, Monsieur Weiffenbach," she answered, never one to stay long with a single manufacturer. "Everyone knows that. That's why we came to you."

He hesitated for a moment, not least because he was bowled over by the sheer force of personality that had come into his office. Laury had not yet managed a single word. "What I can do is to prepare a 134 Super-Luxe rally car for you," Weiffenbach said. "It may be rather expensive."

"I'll pay whatever's necessary," Lucy said. They struck a handshake deal.

Weiffenbach neglected to reveal that Jean François was already ruminating over how to shoehorn the 3.2-liter engine into the chassis of its junior sister, just as the Schells suggested. The reason: Delahaye itself would be fielding competitors in sports-car races the next season.

Great ambitions were afoot.

Part II

4

Crash

R ENÉ DREYFUS SEEMED to be heading for his first win of the season
at the Grand Prix du Comminges on August 14, 1932. Speeding
around the fifteenth, and penultimate, lap in a Bugatti T51, he widened
his forty-second lead over his nearest competitor, Jean-Pierre Wimille.
Then, in a sudden shower, rain poured from an overhanging cloud, pelt-
ing the spectators in the hillside grandstand that overlooked the Pyrenees to the west and soaking the road. The downpour disappeared as
quickly as it came.

Miles away, zooming down the road beside the River Garonne, René
never even knew about the cloudburst. Minutes later, he shot down the
long straight before the right-hand, uphill curve by the grandstand. As
he went into the slick bend at 100 mph, he felt his Bugatti lurch sideways.
The spectators gasped. He twisted the wheel and changed gears, trying
to regain control — all the while fully aware of the horror of what was
happening. For a millisecond, he thought he had regained a purchase on
the road when his left rear wheel struck a bump. Instead, the slight leap
of his tire was high enough for the physics of flight to take effect.

The Bugatti soared upward on its front end and corkscrewed in the
air. Thrown from the cockpit, René bounced across the pavement like
a rock skipped across a pond. His car leveled an acacia tree on the roadside and staggered to a halt in front of the press stand. Dazed, confused,

Dreyfus crashes his Bugatti at the Comminges Grand Prix, 1932

René tried to stand. Officials rushed to his side before he fell, his face covered with blood.

An ambulance hurried him to the local hospital. As he drifted in and out of consciousness, the doctors cut off his overalls and examined him. He suffered several bad lacerations and a severe concussion, but he would live. At one point, he opened his eyes to see Wimille enter his hospital room. René muttered something like, "Did you win?" before passing out again.

When he awoke, his countryman remained in the room. In fact, he was in the neighboring bed, his head wrapped in bandages. The twenty-four-year-old had crashed at the same spot. Neither he nor René knew who finished first until Freddie Zehender entered with a bottle of champagne. He had been in third position, three minutes behind Dreyfus and Wimille, but won nonetheless.

"Thank you very much," Zehender said. "Let's have a drink."

For the next week, René remained in the hospital, reliving the terrifying accident.

Death had always been intertwined with motorsport, but never more so than now, when race cars had attained average speeds of over 120 mph on several courses. Its presence, René once said, "hovered around us." Every driver knew that death might carry any one of them away at any time, whether because of faulty brakes, debris on the track, or a fellow competitor's actions. Traveling the equivalent of 60 yards a second, one could make the smallest mistake and suffer ruin. The ambulances were called "bone collectors" for a reason.

Death was something they understood intimately. They had seen it so often, had come so close to it so many times. They measured how fast to take a corner by how likely it was that they would kill themselves while doing so.

They could not perform if fear was a factor, and so they looked at death not with fear but with something more like respect. One driver likened competing in the face of death or crushing disability as a test "to show the wood from which he was carved." Still, René was shaken by the Comminges accident. He had never come closer to perishing in a race.

Louis Chiron visited him in the hospital. For a split of the winnings, he had supplied the T51 to René and promised to repair it or give him another. "Get back to it" was his advice.

René loved the life of a race car driver. The tinkering with engines, the travel every weekend, the competition, the push to get better with each turn of a circuit and with each and every race.

He enjoyed the fame too: the newspaper headlines, the attention everywhere he went. He was young and flush with cash. Only Maurice kept him from spending it all on custom suits, bar tabs, and gifts for friends — and on ladies. Attractive well-wishers flocked around the pits and at the hotels afterwards; René was no stranger to their charms.

The community of drivers felt almost like family. The best often switched between teams, and a number of independents drove various makes to success, as René had done at AVUS. They traveled together between races in flashy cabriolets. They stayed at the same lavish hotels, dined and drank at the same white-cloth restaurants and tony bars, and even shared the same shirt-maker. After races, there was usually a tuxedoed gala where they huddled together away from the autograph-hunters. During the winter, they skied together in the Alps as they waited to see what the next season would hold. Their girlfriends and wives were often friends.

Among this band of brothers, personalities and high jinks abounded. Prince Bira of Siam traveled with a set of wooden windup toy cars that the drivers used to race in hotel lobbies, and the pipe-smoking Italian Count Carlo Felice Trossi owned a medieval castle with its own aviary. Then there was René's good friend Stanisław Czaykowski, who would end yet another Pernod-fueled bender by knocking on hotel doors at midnight, announcing, "Open in the name of the law!" Once, at a postrace party in a hotel, a mad scramble ensued after Rudi Caracciola lost his mascot, Moritz, a golden long-haired dachshund. When the dog was spotted paddling down the river beside the hotel, René and drivers from every team dove into the water to rescue the poor animal.

They shared a strange kind of friendship. They could enjoy a laugh one night together and the next morning exhaust every effort to beat each other, even if that meant almost running each other off the track. When Nuvolari fell foul of one such incident, he told the driver, when he tried to apologize afterward, "Let's not be sentimental. Motor racing's no game for schoolkids. I'll get my revenge."

Above everything, René loved driving, that feeling behind the wheel when he was in balance with the machine and everything was clear because nothing else in the world existed. Getting back to it was his only choice, despite the dangers that were very real to him in a way they had not been when he was flinging his first Bugattis around Nice.

After leaving the hospital, René entered a couple of more races that season in another T51, but retired from both. Nevertheless, he had done well running as an independent after his break from Maserati, and he finished the year ranked the fifth-best Grand Prix driver. He might not have challenged the new Alfa Romeo P3s, but race after race he often beat Chiron and his Bugatti factory teammates. In the right car, with the right team, he could realize his ambition to be the European Champion.

Late that fall, he was asked to lunch by Bartolomeo Costantini. A flying ace in World War I, Meo Costantini had competed in Bugattis after the war, winning several major races before becoming the team manager for the Molsheim firm. Over lunch, the tall, stately, and often solemn Costantini made his offer to René to join the team. "I'm ready," René said mildly, though he wanted to leap from his chair.

The next day, René signed the contract and met Ettore Bugatti at last. Sporting his trademark brown bowler hat, Le Patron was a man in his early fifties with a round, fleshy face and blue eyes. Most distinctive were his hands; constantly moving his long, elegant fingers, Bugatti reminded René of the conductor of an orchestra. The two men shook on their agreement, and René beamed with delight. He rang his family in Nice, and word of his success spread through the whole Dreyfus clan.

That evening he celebrated with his brother Maurice, who now lived in Paris and worked in the raincoat business. Their overjoyed conversation and clinking of glasses soon brought the whole restaurant in on the great news.

Early in 1933, René moved to a room in the Hotel Heim in Molsheim. To say that the village in Alsace was a company town was to understate the matter; it was more of a fiefdom. The restaurants, grocery stores, pharmacies, and schools were all supported by the fact that in 1909 Ettore Bugatti had decided to convert a dyeworks into an automobile plant. In its first year, after recruiting skilled mechanics, carpenters, and metalsmiths, the firm produced five cars, each a piece of art.

The distinct imprint of Le Patron was visible everywhere at his Molsheim works. Every polished-oak, brass-handled door was secured by a lock etched with MADE BY BUGATTI. Foundries produced the engine blocks and brake drums. A private power plant provided electricity. In on-site workshops, leather was hand-stitched for the seats and glass was blown for windows. Even the tools were customized for particular Bugatti refinements. It was this attention to detail that was Le Patron's polestar—not the pursuit of money.

René spent most days at the factory with Costantini, working on one of the cars. They tested carburetors and different fuel mixtures. They ran engines on the dynamometer and adjusted chassis layouts. Although the Maserati brothers had also allowed him to see behind the curtain, they gave him only a sliver of the engineering education he received from the Bugatti team manager.

In Pau, a provincial French city in the foothills of the Pyrenees, the 1933 Grand Prix season would begin. Shortly before, reports came that President Paul von Hindenburg had named a new Reich Chancellor, and that he was making a stir in the race-car world that would quickly make Bugatti an afterthought.

• • •

On February 11, in Berlin, Adolf Hitler walked into Kaiserdamm's vast "Hall of Honor" to open the Berlin Motor Show. As he stood on a high, well-lit pulpit dressed in a black suit, silence fell.

Only twelve days had passed since he had assumed leadership over Germany. The Nazi Party had celebrated with a torchlight parade through the capital. André François-Poncet, the French ambassador, described the striking scene in his memoir, *The Fateful Years:* "In massive columns . . . they emerged from the depths of the Tiergarten and passed under the triumphal arch of the Brandenburg Gate. The torches they brandished formed a river of fire." The brownshirted, jackbooted tide marched past the French embassy, then down Wilhelmstrasse, raising its voice as it crossed by the winged palace of the Reichspräsident. Watching from one window was the aged General Hindenburg, a gnarled grip on his cane. In the next stood Hitler, the parade of light flickering in his eyes.

The new chancellor had moved swiftly to secure his rule. He called for new Reichstag elections, purged state offices of political opponents, arrested thousands, green-lit attacks on Jews, commandeered radio stations, and recruited business leaders to finance his election campaign. Feeding off the drummed-up fear of communist violence, he aimed to suspend a host of personal liberties guaranteed under the Weimar Constitution, including the freedom of the press, the right to assemble, and the need for a warrant to search someone's house. To cap this flurry of activity, he decided to deliver one of his first major public speeches as chancellor to promote the automobile industry. Ideally suited to introducing Hitler was Manfred von Brauchitsch, with his tall, broad-shouldered frame, sweep of fair hair, and easy dashing manner. As requested by Mercedes, Rudi Caracciola attended as well.

Voice quivering with a passion amplified by loudspeakers, Hitler declared his intention to slash the taxes and regulations that were suppressing the motor industry, to align its growth and development with aviation, to build a nationwide highway system, and to dominate international motorsport. The captains of the industry, including Wilhelm Kissel of Daimler-Benz, hung on his every word. "These momentous tasks are also part of the program for the reconstruction of the German economy!" the new chancellor exclaimed.

With these rousing words, Hitler declared the motor show open.

Hitler and his "Cavalry of the Future," including (first row, left to right)
Bernd Rosemeyer, Rudi Caracciola, Adolf Hitler, Hans Stuck,
Adolf Hühnlein (in uniform), and Ernst Henne; (second row far left)
Manfred von Brauchitsch, 1935

Among those accompanying him around the exhibition halls afterward
was Adolf Hühnlein, whom Hitler had personally tasked to command
the National Socialist Motor Corps (NSKK). Hitler saw revitalizing the
automobile industry not only as a key pillar to deliver on his promise
of a restored Germany but as a factor in strengthening the Reich for future war. The purpose of the NSKK was to train a legion of men in motor skills, establishing the foundation of a mechanized army. For now,
as Hühnlein often stated, they were merely part of the "defensive force
of the nation."

The promises made by Hitler at the Berlin Motor Show significantly
brightened Daimler-Benz's prospects for sales, and every opportunity
now needed to be leveraged to their advantage. When the board gathered in their Stuttgart headquarters in March, Kissel focused on how to
continue to exploit these aims and build on their already close relation-

ship with Hitler. One of the first agenda items was the dismissal of a Daimler-Benz lawyer for making anti-Nazi political statements.

Over the past decade, the closer Hitler had come to power, the closer Daimler-Benz drew itself to him. Jakob Werlin, head of the company's Munich dealership, had long fostered a friendship with the Nazi leader. Before the failed Putsch in 1923, Werlin had sold Hitler his first Mercedes, a symbol, as historian Eberhard Reuss wrote, of "force, strength, power, and superiority." The dealer continued to provide discounted vehicles for Hitler's cross-country political campaigns and became a self-declared "personal confidant" of the fascist leader.

During his rise to power, Daimler-Benz advertised in the Nazi Party's newspapers. The company curried favor with Hitler by sending their most successful and famous drivers to visit him. Board members raised funds for the Nazis and solicited the party's top officials to engage the firm in the event of the rearmament of the military. Using Werlin as a go-between, Kissel ensured that this rising political force was aware that his company had "no occasion to diminish the attention which we have until now afforded Herr Hitler and his friends; he will be able to rely on us in the future, as in the past."

Now that Hitler had ascended to the position of chancellor, Daimler-Benz needed him far more than he needed them. With car and truck sales plummeting by half from their record high in 1928, Kissel believed that government support was the best path out of the crisis. Moves by the Nazis to crush the trade unions would help the Daimler-Benz bottom line. More critically, orders for heavy trucks were rising, and there was a clear signal now that Hitler intended to reinvigorate the German military in direct contravention of the Versailles Treaty. Prototypes for airplane engines and tanks had already been ordered. If they moved toward full-scale production, it would be a huge boon to filling the excess capacity at many of their plants and to improving profits.

A return to racing would be a crucial next step: wins on the international stage would provide a marketing bonanza, particularly when it came to growing the export market. Kissel knew that engineering race cars for the new 1934 formula—and financing a team of drivers and staff—would be expensive. The new formula decided by the AIACR aimed to limit speeds in excess of 140 mph and stipulated a maximum weight limit of 750 kilograms without driver, fuel, oil, water, and tires. Kissel's staff calculated that the company would need to spend at least 1 million marks a year ($7.75 million in today's dollars) to reign supreme in the Grand Prix. Only through state funding could this ambition be realized. Already the Mercedes driver Brauchitsch had personally pitched Hitler to provide such for the company. "You will receive the money the moment I rise to power," Hitler had promised.

Besides having to wait for these monies to begin to flow so they could build "Reich racing cars," there was another problem. Auto Union, the result of a recent merger of four depression-hobbled German automobile manufacturers (Horch, Audi, DKW, and Wanderer), was also soliciting state sponsorship of its racing program. They had already contracted Ferdinand Porsche to design a car under the new formula. The brilliant engineer had launched his own charm offensive on Hitler, complimenting him in a letter after the motor show for his "profound speech" and his desire to "place our skill and determination at the disposal of the German people."

At the March 10 board meeting, Kissel presented the draft of a letter to Hitler, suggesting to the "Highly Esteemed Herr Reich Chancellor" that state funding was needed to compete aggressively in the Grand Prix and that Daimler-Benz alone should receive it. "Since in the course of the company's history," Kissel concluded the letter, "our marque has contributed frequently and significantly to the respect paid to Germany at sporting events, we would dedicate all our skill and knowledge to this and would deem it an honor if we were enabled to represent the German flag in the sport of the future."

They soon received their answer. The Reich would fund the 1 million marks. But the sum would be split between Daimler-Benz and Auto Union. Further, Korpsführer Hühnlein would be charged with overseeing their Grand Prix efforts.

Although Daimler-Benz received only half of what they thought they needed to support a team, they knew they had no choice but to accept it—and to do their best to bring a winning race car to the new formula. Gaining the new chancellor's favor promised to be worth the expense.

Kissel motioned the board to begin recruiting drivers for their new racing car. Manfred von Brauchitsch was eager. Negotiations were ongoing with veteran Mercedes driver Hans Stuck (who eventually decided to race for Auto Union). The big question was Rudi Caracciola. He was committed to his own team for 1933, but Alfred Neubauer had secured his assurance that he would not sign with a factory for the following year without first consulting Mercedes. It was paramount that Germany's best driver be ready to lead their team to victory in the new formula year.

On Thursday, April 20, 1933, Rudi was blazing around the streets of Monaco in a 2.3-liter Alfa Romeo painted white with a blue stripe running down its side. A short distance behind followed Louis Chiron, his car blue with a white stripe (a nod to their French-German partnership). Both carried the symbol of a pair of backward-facing Cs, which represented their newly formed team, Scuderia CC.

Rudi Caracciola and Louis Chiron, partners in Scuderia CC

The previous fall, Rudy and Charly Caracciola had hosted Louis and his longtime girlfriend, Alice "Baby" Hoffmann, at their Arosa chalet in the Swiss Alps. American-born, the daughter of jet-setting parents, Baby had been married to Alfred Hoffmann, heir to the Swiss pharmaceutical giant Hoffmann–La Roche, before falling for motorsport, then for Louis. She spoke five languages fluently, and as one contemporary described her, she inhabited the personality of each one: "French charm,

German perseverance, American business sense, Swedish sex appeal, and Italian temperament."

The two couples already spent a lot of time together. Charly joked to other visitors that Baby, whose framed photograph stood on the mantel, was "Rudi's silent love." Despite successful seasons, both had been let go from their teams—Louis because of personal conflict with Bugatti, and Rudi because Alfa Romeo had decided to abandon its works team. "You know," said The Old Fox, "why should we always win the prizes for other people? It would be much smarter to start our own firm." Thus, "Scuderia CC," after their two initials, was born between two of the Grand Prix's greatest drivers.

As Rudi and Louis shot past the Monaco pits on their twenty-fourth practice lap, Baby and Charly clocked their times. In a Grand Prix first, the starting grid on the day of the race would be determined by the best time achieved during the three days of practice. On the initial day, Rudi and Louis focused more on testing out their Alfa Romeos than anything else. Louis had never driven the make before, but under a sky pocked with black clouds, the two drivers managed to clock the fastest laps that morning: two minutes, three seconds. Tazio Nuvolari and his Italian archrival Achille Varzi were one second slower. René Dreyfus was off the pace by three seconds.

By their last run of the morning, the sky was clear, but a curtain of mist hung over the bay. Rudi continued to lead down the corkscrew turns to the seafront, with Louis on his tail. He was amazed at how quickly Louis had gotten a feel for the Italian race car. The two shot through the tunnel. Back in the sunlight, Rudi zipped through the chicane, then accelerated down the straight toward the left-hand turn at Tabac Corner. A glance in his mirror showed Louis nowhere in sight.

Rudi braked slightly while looking in his rearview mirror to see where his teammate had gone. Suddenly, his Alfa went into a skid. Only one of the front brakes had engaged, he thought, as his car swept at 70 mph toward the stone parapet that separated the promenade from the sea.

Time slowed to a crawl. He cranked down through the gears. Calculating that he was more likely to survive a smash into Tabac Corner than a leap into the water, he steered away from the parapet. Hands tight on the wheel, he tried to regain control.

He was moving too fast.

His car snaked left, then right, on the road.

The stone steps came closer, and closer.

At last Rudi regained control, but it was too late. There was nothing to do.

The Alfa Romeo struck the wall first by the right wheel, then the whole side panel. The white metal body collapsed against the stone. Then the car propelled sideways for a few dozen feet before coming to a juddering stop.

Rudi was stunned, but he thought he was okay. He did not feel the streams of blood coursing from his temple, nor realize that his thighbone was completely crushed. He wanted only to be free of the car that seemed to have molded itself around his body. With strength born of shock, he wrested himself out of his seat.

Several people were dashing down the steps from the upper road toward him. He was fine, Rudi thought. No trouble here other than a wrecked car. Behind him there was a squealing stop. Louis jumped out of his own Alfa.

Rudi tried to take a step forward, and an explosion of pain overwhelmed him. His right leg gave out, and only Louis's arrival at his side kept him from collapsing onto the road.

In the pits, Charly and Baby waited for the two Alfas to finish their twenty-fifth lap. They should have come by now. A shroud of worry that something terrible had happened fell over them.

Rudi was carried away from the track in a simple wooden chair taken from a café. He sat upright in the chair in a state of shock. Blood ran into his eyes. The whine of race cars circling the track was deafening. Finally the ambulance arrived. Its crew placed Rudi on a stretcher and jostled him inside. Each bump and turn through the streets of Monte Carlo sent ripples of pain through his leg. Something was very wrong with it. Rudi dared not ask what.

At the hospital, he was carted first into the X-ray lab, then into the surgery ward. Waiting for the doctor, he stared through the high windows at the treetops waving in the wind. Everything in the room around him was sterile white and glass. The pain he had felt earlier was only a fraction of the agony that swallowed him now. His face was lacerated in several places, sweat beaded on his forehead, and the grim lock of his jaw spoke of his suffering.

Finally, a Dr. Trentini arrived. He was short and sallow-skinned. Rudi disliked him instinctively. Neither spoke more than a few phrases in the other's language, and they had trouble conversing. Rudi just wanted him to get on with setting what was assuredly a broken leg. He wondered what was the delay.

Dr. Trentini and his assistant stood by the window, examining the X-rays, when Charly came into the room with Louis and Baby. "Tell them to pull my leg as hard as they can," Rudi urged. He had known other drivers to come out of such injuries with one leg shorter than another, which would have been unacceptable to him.

Dr. Trentini drew Charly from the room and raised the X-ray into the light. "Look, madame: The femur and the entire tibia bone are completely smashed. Your husband will never be able to drive again."

Charly almost fainted. Baby did when her friend told her what the doctor had said.

Three days later, on Sunday afternoon, Rudi was lying in his hospital bed, his right leg in an ill-shaped plaster cast up to the hip. Charly sat beside him, and flowers from well-wishers covered every available space. They were listening to a radio broadcast of the race, trying to decipher the commentary in French about the feverish contest between Nuvolari and Varzi.

In the final lap, near the finish, the engine of Nuvolari's Alfa Romeo engine died and caught on fire. He leaped out and tried to push it toward the finish, enveloped in billowing black smoke. Driving a Bugatti, Varzi won easily, followed by Mario Borzacchini, with René Dreyfus in third. Louis Chiron was a distant fourth.

Despite everything, including the hurried consultation from an Italian specialist who had saved his leg from the saw, Rudi was stunned that he had not recovered in time. *He belonged in the race. That was his place in the world.*

The plain fact was that his legs would never be a balanced pair again. He would have a permanent limp and was likely to only be able to take a few hobbled steps at a time. Racing looked like but a glory of the past.

• • •

At the Monte Carlo Casino, René and his Bugatti teammates, including Varzi, dined on lobster and celebrated their pooled winnings. A reporter broke into their revelry, but they said nothing about Rudi Caracciola. It was always sad to see one of their brethren crash, but dwelling on such calamities only heightened the chance that they might suffer their own doubts. When the reporter asked what they did to keep in shape, Varzi hurried off, a cigarette angled on his lips.

"Do not ask him about winter sports," René said dryly. "He's too afraid to break his arms and legs."

The table erupted in laughter, and the night carried on.

After his third-place finish at Monaco, René returned to the track at the Belgian Grand Prix. Again he placed third. At Dieppe and Nice, second. First place eluded him, chiefly because the Bugattis in the Pur Sang (thoroughbred) stable were simply behind the times when compared to the Alfa Romeos and Maseratis.

Between competing and spending time at Molsheim, René fell in love. Nicknamed "Chou-Chou" (in French, the equivalent of "sweetheart"), Gilberte Miraton was one of a coterie of wealthy young women who frequented the racetracks and modeled the latest fashions while competing in Concours d'Élégance events—Chou-Chou with her grand white Delage.

Chou-Chou and René crossed paths in early 1933. Vivacious, funny, and whip-smart, she was impossible to overlook. She was short and dark-haired and had the kind of presence that drew the attention of everyone in the room. René had dated other women, but she was the first to hold his interest. Whenever possible, he visited her at her home in Châtel-Guyon, a spa town near Vichy in central France, or they met at races.

As his first season with Bugatti neared its end, René was happy and content. He did not like what he was hearing about the political situation in Germany, but any effect it would have on his life seemed distant. He had seen Fascists and their work firsthand while living in Mussolini's Italy. One evening in May 1931, he and the Maserati brothers were sitting on the terrace of the Caffè San Pietro in Bologna when a crowd of theatergoers poured into the streets from the Teatro Comunale di Bologna. The journalist Corrado Filippini, who was also sitting at the Maserati table, went off to learn what had happened. He returned with a disturbing report. The conductor Arturo Toscanini had refused the order to start the evening with the official hymn of the Italian National Fascist Party, "Giovinezza," as was now expected at public performances. Irked by his act of defiance, one of Mussolini's ministers who was present ordered a gang of toughs to attack the aged maestro when he left the theater. They surrounded Toscanini and hit him until he fell. René had lived in Italy for long enough to experience the strutting Fascists but had believed them to be harmless buffoons. The Toscanini beating made him see the uniformed blackshirts in a very different light.

The Nazi variety of the species looked to be much worse, particularly toward Jewish people. France had its own deep threads of anti-Semitism that pervaded its culture, most pointedly seen in the trial of French army captain Alfred Dreyfus, who, against clear evidence, was accused of selling military secrets to Germany and convicted of treason in 1895. The "Dreyfus Affair" was far from ancient history by the 1930s, and prejudice against the small population of Jews remained the standard. A Jesuit in France described it as "anti-Semitism of principle —latent and quite general." Historian Eugen Weber added: "One did not have to think ill of them, to prefer to avoid them, and, of course, not to want one's offspring to marry one." Jews may have had French passports, but they were viewed as "other" by many of their compatriots.

Despite this underlying anti-Semitism, and even though he shared the surname of the most famous Jew in France (they were not related), René had rarely faced much prejudice. In his upbringing, he was insulated by his tight-knit community in Nice. Although his father came from a conservative Jewish family, he did not attend synagogue. René was never bar mitzvahed. If anything, he had a closer association with his mother's lapsed Catholicism by virtue of the fact that he and his siblings mostly spent time with her side of the family. Neither religion made any impression on him, and he might have called himself an atheist if he had given any thought to it, but he hadn't. As a driver, his heritage had never affected his prospects, nor could he imagine it ever would.

"They know there is a menace," one French essayist said of his fellow Jews at the time, "but they bury heads in the sand." René was in this camp, his focus only on motorsport.

If René perceived any danger from Germany, it was on the race track, where disaster soon signaled the terrible cost of the endless quest for faster cars. On September 10, he and Costantini were eating lunch at the Molsheim canteen and listening on the radio to the heats before the main race of the Monza Grand Prix.

The Autodromo Nazionale Monza, north of Milan, was a very fast oval track that already had the reputation as "The Circuit of Death."

Suddenly, the announcer went quiet. René thought the reception had cut out; it had been static at best up to that point. Costantini was certain there had been an accident. A short while later, a phone call from Monza brought news of a crash on the autodrome's south curve. Giuseppe Campari had lost control of his Alfa Romeo P3, swerved left, and launched over the banked curve. The much-loved Italian was killed instantly.

Attempting to avoid Campari, Mario Borzacchini hit the brake pedals hard. However, his mechanics had removed his front brakes, and they had fitted his Maserati with smooth-tread tires to increase his speed. He never had a chance. His car slewed over the concrete retaining wall, and he was thrown into a tree, which broke his spine. He died in the hospital.

Despite protests to stop the heats, the final one was run that afternoon. In the eighth lap, Count Stanisław Czaykowski was in the lead when he flipped over the embankment and was trapped under his burning Bugatti. Some spectators tried to drag him from the car, but they were too late to save him.

The next day, Benito Mussolini laid flowers at a hastily built memorial at Monza for the three Grand Prix champions. The Italian leader was a fervent fan of sport. In 1928, a bank controlled by his Fascist government had taken over the struggling Milanese automobile manufacturer Alfa Romeo from its founder, Nicola Romeo. Mussolini owned his own fleet of Alfas and subsidized their racing in part through contracts to build aero engines and military equipment. He established the model of state support that Hitler was keen to follow.

If anyone had any questions about who was really in charge of the Alfa Romeo team, events at Monza two years before had answered them. During practice, Alfa Romeo driver Luigi Arcangeli died when his car shot off the track and smashed into a tree. His teammates wanted to withdraw in homage to "The Lion of Romagna," but Mussolini squashed the act of sentiment: "Start—and win!" he ordered his drivers.

Now three more lives had been lost in this mad pursuit of victory, and their black-veiled widows suffered through the funeral scene, which journalists described as the passing of the old guard and the worst catastrophe in racing history.

René had been closest to Count Czaykowski, and his death struck him deeply. Czaykowski was a merry, generous *bon vivant* who golfed as well as he drove, and they had often traveled together on the road. The following week, René attended his friend's funeral. The family had placed his car's charred steering wheel, wrapped in French and Italian flags, on the coffin. The sight spooked René.

Also listening to the race that day, his right leg bound in plaster, was Rudi Caracciola. He was in his room at the Rizzoli Orthopedic Institute, in Bologna, which was housed in an ancient hillside monastery south of the city. Its director, Dr. Vittorio Putti, was the preeminent Italian surgeon in his field.

Over the previous five months, Rudi had mostly lain in bed, playing cards with Charly or gazing out his window. Every time there was a race, he listened to it on the radio. Midseason, Louis Chiron had joined the team started by Enzo Ferrari. Since Alfa Romeo had abandoned its own factory team, it had commissioned Scuderia Ferrari to field its cars. Piloting the *monoposto* P3—emblazoned with the soon-to-be-famous prancing-horse badge—Chiron won the Spanish Grand Prix and two other races. Missing out on such an opportunity was torturous to Rudi. He wanted nothing else but to drive again.

His progress remained unclear. After each inspection of Rudi's leg, Dr. Putti would only declare, "Well, it'll be all right . . ."

At the end of September, Putti cut off Rudi's plaster cast, and two nurses rolled him into the X-ray lab. He hoped that the long struggle might soon be over, but Putti declared later that evening that more time in plaster would be needed. In late October, he removed this second cast. Rudi attempted to walk on crutches, but the effort was too much. X-rays revealed that the cartilage in his right leg was healing improperly. Having made sure that Charly was in the room, Putti recommended surgery. Rudi refused. He was convinced that he just needed to get back on his feet. The thought of more surgery and more months in plaster was intolerable.

"You won't be able to drive anyhow!" Charly exclaimed. Her declaration stunned Rudi. Since Monaco, nobody had dared say that his smashed thigh bone was career-ending. Although his right leg was now two inches shorter than his left, Rudi was certain that he could overcome this challenge. When Putti advised that walking was likely ambition enough, Rudi felt a coldness spread inside his body.

Charly tried to tell him that there were other things in life, that they could be happy without racing. He silenced her. "No surgery," he told Putti. However, he consented to returning his leg to the plaster tomb on the condition that he could immediately leave the Institute.

They moved to a friend's house in Lugano, Switzerland. Between periods of rest on the terrace overlooking the lake, Rudi tried to spend more and more time on his feet. Despite using crutches, each swing of his leg sent a shot of pain through his hip. Charly walked beside him in case he stumbled.

In mid-November, Neubauer paid him a visit. Rudi hid his plaster cast under loose trousers. The Mercedes team manager gave him a great hug, then they retired to the terrace. Rudi sensed that his every move—and facial expression—was under inspection.

As always, Neubauer got straight to the point. Hitler was supporting Daimler-Benz in developing a new race car for the 1934 season. Brauchitsch and Italian champion Luigi Fagioli had already signed on to the team. Neubauer wanted to know if Rudi could drive again. They needed him.

"Of course, I can," Rudi said, before brazenly asking about the contract. Neubauer waved away the question. Rudi needed to come to Stuttgart to discuss that—perhaps in January. The two then spent a pleasant afternoon together. Little was said about the car the company was developing, or about its sponsor, who had already firmed up his grip on absolute power. That was not their world, not their concern. They only hoped that support for Mercedes would continue.

After Neubauer returned to Germany, Rudi learned through a friend that the team manager had reported to Kissel that there was little chance Rudi would compete again. Writing him off, they were looking for younger, fitter drivers.

Doctors removed the plaster cast in December. Every day, Rudi tried to walk a little farther, cane in one hand, Charly holding the other. He was gathering strength, but his uneven legs made for an awkward gait, and the pain in his hip never faded.

In early January 1934, he and Charly traveled to Stuttgart. She was resigned to the idea of him racing again. Rudi had pushed for a meeting with Kissel to tell him he was healthy enough to drive. The Daimler-Benz CEO was unconvinced but promised that Neubauer would sit down with him later that evening to discuss it.

Neubauer came directly to Rudi's room at the Graf Zeppelin Hotel. "Fit and well again?" he asked. As he had done with Kissel, Rudi masked his limp and gritted his teeth through the strain. He swore he was up to driving again. Neubauer reminded him of the strength needed to brake and accelerate hundreds of times during a race. "What guarantee is there that you won't crack up in the middle?"

Charly lost her temper, suggesting that their new car might "crack up" first. At last, Neubauer offered Rudi a compromise. In May, when the car was ready for testing, Rudi would have to participate in a practice run. If he passed muster, he could be on the team.

The next day, Neubauer brought him to the Untertürkheim plant to show him the new car. They passed a series of huge workshops and arrived at a small building surrounded by a high, barbed-wire fence. A guard checked their identification before allowing them through the gate. Everything about their work was top secret, Neubauer warned. Inside the workshop, Rudi got his first look at the engine and the overall design of their new Grand Prix car: the W25.

The theory behind the 1934 formula was straightforward: if factory teams wanted more powerful engines, the chassis and other gear needed to be heavier to hold them to the road; with weight limited to 750 kilograms, speed would therefore also be limited. Dr. Hans Nibel, head of the company's design department, figured differently. In his interpretation, the formula basically allowed an unlimited engine size if one could design a lightweight chassis and running gear with the ability to corner, brake, steer, and hold to the road while traveling at very high speeds. Achieving such driving performance was no small matter. As Ferrari historian Brock Yates wrote, up until then, "cars, even the somewhat advanced P3, were merely crude four-wheeled platforms upon which to plant [huge] engines."

The prototype Mercedes engine, a supercharged 3.3-liter straight-eight, which Rudi saw mounted on a dynamometer, was an absolute thoroughbred. Although not a revolutionary design, it benefited from ultraprecise construction and a host of improvements, allowing for horsepower measurements 50 percent greater than the Alfa P3.

Nibel matched the engine with a refined platform built of light alloys wherever possible. The W25 was a single-seater — new territory for Mercedes, as was its use of hydraulic brakes. It featured front- and rear-wheel independent suspension. They also advanced design by combining the gearbox and differential (which provided power from the crankshaft to the rear wheels) into a single unit. This provided better weight distribution and allowed the driver to sit lower in the cockpit. The W25 promised to ride balanced and tight to the ground. If everything was tuned correctly, it would be the fastest race car to ever grace the Grand Prix.

Motivated more than ever to recover, Rudi moved back to Arosa. During the bright winter days, he bathed his leg in sunshine. In the evenings, he ambled out for a walk with Charly, venturing farther each time. He accepted that he would never again move without pain. The real question was his ability to endure a 500-mile race in the tight W25 cockpit.

On February 2, Charly headed off for a day's skiing with some friends. She was reluctant to go, but Rudi convinced her that she deserved at least a day's vacation after ten months of nursing him. Later that afternoon, he hobbled down the snowbound hill to meet her at the train station. Neither she nor any of her ski companions showed up at the appointed hour. He returned home.

The sun set, and there continued to be no sign of Charly. Rudi sat in the chalet, the lights out, the better to see her coming up the road. At 10:00 p.m., the guide who had led her tour approached the door. One look at him and Rudi went white. Words followed: There had been an avalanche . . . Charly must have seen it; instead of dodging its approach, she looked to lean into its advance . . . her friends believed she might have had a heart attack; she was known to have a weak one, didn't he know? . . . Charly was dead.

Without her and without racing, there was nothing left in his life.

5

The One Thing

A T THE DELAHAYE factory, hammers pealed, machine tools grumbled, compressed air hissed. The smell of ground metal, oil, grease, lacquer, and the heavy effort of a team of men permeated the air. Past them strode Lucy Schell, once again headed toward the office of Charles Weiffenbach. She was incensed. Not only had his company missed the delivery date of her Delahaye Type 134, preventing her from competing in the Monte Carlo Rally in January 1934, but he had then supplied the very model she proposed be built for his factory driver, Albert Perrot, to participate in the Rally himself. Perrot had placed a dismal 24th out of 114 participants.

It was an experiment only, Monsieur Charles said, trying to pacify her. Then he further stoked her anger by revealing that he was supplying a pair of that same "experiment," 138 Specials, to two of her rivals in the all-female Saint-Raphaël Rally, Mademoiselle Gonnot and Madame Nenot.

The two women were longtime Delahaye customers, Monsieur Charles said. Surely she must understand.

Lucy did not, and to prove her point, she crushed the two 138 Specials in the Saint-Raphaël Rally, driving her rather less powerful Delahaye 134 to a fourth-place overall finish.

While Lucy fought her battle with Monsieur Charles, the daily newspaper headlines chronicled a France spiraling into chaos. The government was a shambles, with prime ministers and their cabinets rising and falling in months rather than the usual years. Shortly before Saint-Raphaël, a grab bag of right-wing forces clashed with police on the Place de la Concorde. Some believed that a Fascist putsch might be at hand. In crushing the protest, fifteen people were killed and hundreds injured. "It appears that France was failing to adjust her social and political system to meet the demands of the modern world," remarked the *New York Herald Tribune.*

Insulated by her fortune, Lucy could focus on racing, but she worried about France — and its strength among nations. Although neither wholly French nor American, she had spent much of her adult life living in and around Paris. She had served there in World War I. Her sons had been born in France, and like her, they spoke the language with as much ease as English. Lucy believed that France must find its footing against an emboldened Fascist Germany, but it was even beyond her sturdy self-belief to think that she could have a role. Her energies were instead bent toward competing on an equal level with the men who dominated the sport she loved.

While women in America had already won the right to vote, they remained disenfranchised in France. The emancipated flapper, who wore her hair in a bob, skirt above the knee, and did as she liked, was largely a myth. Promiscuity was likened to prostitution, contraception was a rarity, and abortions were forbidden. Women were largely expected to produce children, keep house, and do as their husband or parents told them. Well-to-do women were certainly not expected to work for money, and any labor they did was to be performed as *dames de charité* (ladies of charity). In her life, every day, Lucy had to fight this model of what she ought to be.

In motorsport, Lucy was up against an even more male-dominated world, one that was riddled with sexism from its earliest days. One of its pioneers, Camille du Gast, was banned from certain events for "feminine excitability." Organizers spent an inordinate amount of time thinking about what these early female drivers should wear (lest, ridiculously, their skirts blow up over their heads), rather than how well they were competing. By the time Lucy first got involved in the sport, the concept of women racing cars — not to mention driving at all — had barely improved. According to *La Vie Automobile,* women were "weak and delicate by definition" and therefore ill suited to the "muscular efforts" needed to start a car, change the tires, brake, or steer. Another editorialist claimed that women cared little about an automobile's power or handling abilities; they were "only attentive to the aesthetic factor." Speed queens like Elisabeth Junek, who was a specialist of the Targa Florio, the Sicilian open-road endurance race that was one of the toughest in the world, revealed these claims to be ridiculous. But many races continued to forbid women outright. Others only allowed them as co-drivers. They were a rarity in Grand Prix events, and factory teams refused to let them join their ranks.

As for their fellow male race-car drivers, they seldom gave women the credit they deserved. One driver joked, "They chase after us on the track; we chase after them off it." When passing female drivers in a race, Louis Chiron liked to blow them kisses. To buck this pervasive prejudice that she experienced in every room she entered — and in every competition she signed up for — Lucy had to be supremely competent and forceful in what she wanted.

After the Paris–Saint-Raphaël, Lucy visited rue du Banquier again. This time she was unequivocal. She wanted a 138 Special of her own for the Paris–Nice Rally so that she could prove herself to be the "best Delahaye rally driver in the land" — man or woman. Weiffenbach agreed, not least because he wanted to use her wins to advertise his new line of cars. There was another factor as well: he liked Lucy Schell. She was every bit as pertinacious, combative, and perservering as he considered himself to be.

On Saturday, March 24, Lucy arrived for the start of the Paris–Nice Rally. She and her five fellow speed queens (out of a total field of forty) posed arm in arm in front of their cars for photographers. Chapeau tilted on the side of her head, black fitted jacket zipped to her throat, and wearing heels, Lucy looked like she had just stepped off the Champs-Élysées after shopping for the latest spring fashions. She then changed into her racing overalls.

The Paris–Nice Rally had the qualities of a decathlon compared to the knockdown, dragged-out marathon of the Monte Carlo Rally. First run in 1922, it served to spotlight, over a number of events, the speed, stamina, flexibility, braking, road-handling, and acceleration abilities of the entrant cars — and their drivers' ability to show them at their best.

Lucy took off on the first stage through a morning fog. The sky cleared a half hour later, and she kept a steady pace south toward Marseille, 750 kilometers away. There were some spells of rain en route, but Lucy finished well within the allotted time as dusk settled over the southern port city.

The next day she ranked third best in the kilometer-acceleration test up the hill of Marseille's Boulevard Michelet. On the Monday, she ran the 200 kilometers to Nice, again right on schedule. In the afternoon, and the next day, she was eighth in the 500-meter flying-start contest and performed well in the braking and steering tests.

For the final stage of the Paris–Nice Rally, the La Turbie hill climb, Lucy barreled away from the start, uncowed by the first sharp turn where an etched stone marker reminded all drivers of the nature of the race. There, on April 1, 1903, Count Eliot Zborowski, a glamorous pioneer of racing, snagged his cuff link on the hand throttle and, unable to slow, failed to make the turn. His Mercedes slammed into the rough-hewn wall. Such was the impact of his flying body against the stone, one account stated, that "his head vanished into his chest cavity."

After the turn that claimed Zborowski, Lucy sped up one of the steepest sections of the course. Bordered by sunflower-colored villas and palm trees, the road headed north, away from the coast, in a series of soft S-bends and long straights that allowed her to push the 138 Special almost to its maximum. The climb steepened again as she headed east toward the finish across from Èze, a town perched like an eagle's nest atop a peak between the Grande Corniche and the coastline. There was not an instant to enjoy the sweeping view as she swept onward. She finished the serpentine course with the tenth-best time, five minutes and 26.6 seconds.

René Dreyfus was a contestant in the hill race alone. He broke the course record with a time of three minutes and forty-five seconds, reclaiming his title as "The King of La Turbie" from Jean-Pierre Wimille. He won in a four-wheel-drive, 4.9-liter Bugatti, which one writer said had "the demeanor of a barroom fighter."

At the Automobile Club de Nice the following night, the race's organizers celebrated the winners. Lucy finished eighth in the overall ranking, first in her engine class, and first again among the female drivers, earning her the Coupe des Dames. Her "admirable virtuosity" as a driver was lauded equally with her "perfect" 138 Special. Standing together, she and René lifted their trophies at the front of the dining room.

The two barely knew each other except by reputation. They moved in the same motorsport circles but had never competed in the same race before. René knew more about her sons, who always seemed to be hanging out at the sides of racetracks or in the pits, their mother gaining them entrance where they would otherwise have been shooed away.

Between the two drivers, there was much to discuss with regard to the pitfalls of La Turbie and the challenges of a contest solely against the clock. No doubt Lucy regaled René about her 138 Delahaye, and he about the Bugatti, whose steering was so tough that he felt like his arms might have broken off midclimb. Neither knew then that they would one day work together to share the victor's stage in a much greater arena: the Grand Prix. For now, Lucy was glad that she finally had her perfect Delahaye.

Six weeks later, on May 8, Jean François's latest creation rolled onto the Montlhéry autodrome, the Delahaye name printed in block letters on its curved hood. The young engineer had essentially taken the 138 Special, souped up its 3.2-liter engine with high-performance pistons and other improvements, and had it bodied in a single-seater, aerodynamic aluminum shell that resembled the fuselage of a fighter plane, complete with bubble-roofed cockpit. The body, designed by the noted *carrossier* Joseph Figoni, was painted with the colors of the French tricolor: red, white, and blue. Monsieur Charles had ordered it to be built specifically to break a record and to garner headlines for the company.

A giant concrete bowl set on a lush hilltop on the road to Orléans, fifteen miles from Paris, Montlhéry was a favorite site for record attempts. Cars could race around the oval track day and night, the screaming of their engines insulated by the surrounding countryside. Constructed in 1924, the autodrome was financed by a French industrialist, Alexandre Lamblin, who also owned the popular *L'Aero-Sports* newspaper. When a road course was added a year later, Lamblin made sure the location went on to host the French Grand Prix.

It was on the autodrome alone that Delahaye intended to wrest the World 48 Hours Record from the current holder, Renault, whose drivers had managed, the previous month, a distance of 8,037 kilometers

(4,993 miles, the rough equivalent of driving from Los Angeles, California, to Charleston, South Carolina, and back) during the allotted time, posting an average speed of 167 kph (104 mph). At 4:00 p.m., Perrot set off. François had secured a forty-gallon gas tank in the car's truncated conical tail so that they would only have to stop every four hours to refuel and switch pilots. Marcel Dhôme and Armand Girod were lined up to take part in the attempt.

The 1934 world-record breaker Delahaye 138

Twelve hours later, the car was smoothly circling the autodrome at an average of 183.7 kph and had covered a distance of 2,204 kilometers, breaking a national class record for that time period. The engine was running at an untaxing 3,800 rpm. Fifteen hours into the attempt, some bolts on the large fuel tank had loosened, causing a leak. François was untroubled. The drivers could rely on the auxiliary tank, though it would now require hourly pit stops. Soon enough, more national records in the 3,000–5,000 kilometer range fell to Delahaye. Then the World 24 Hours Records succumbed as well.

At lunchtime on May 10, the track car was continuing its inexorable march, round and round, records falling with easy regularity. Monsieur Charles and a host of Delahaye staff were celebrating the string of successes in La Potinière, the café behind the grandstand, when a company timekeeper threaded through the maze of round tables to reach them. Given the current pace, he told Weiffenbach, they had a shot at the 10,000-kilometer record as well. *Go for it,* was the response.

A weary, hood-eyed François watched his track car sail past at 4:00 p.m., the World 48 Hours Record theirs, covering 8,464 kilometers with an average speed of 176 kph, comfortably beating Renault. Past midnight, the trio of drivers finished the distance of 10,000 kilometers, breaking yet another world record. In sum, Delahaye claimed eleven national class records and four world records with its "son-of-a-truck" dynamo of endurance.

Jean François had clearly stepped out of the shadow of Amédée Varlet, the firm's long-standing engineer. François had an unlikely background for an automobile savant. Raised in Revel, a French town noted for cabinetmaking, and educated at a university famous for producing Catholic theologians, he nonetheless led the design department at Beck Automotive in Lyon by his thirtieth birthday. His advanced prototypes failed to save that company from bankruptcy, but they provided him with a promising résumé for a job at Delahaye. When François arrived at the rue du Banquier in the unbodied chassis of his Beck design, Weiffenbach hired him immediately. Over the next decade, he largely played the dutiful assistant to the staid ideas of Varlet.

No more. The day after Delahaye's record-breaking success, *L'Auto* featured his role and remarked that the "old French firm, whose construction has always been irreproachable, has returned to the sports field to enchant us."

At a party to celebrate the achievement, also attended by Lucy Schell, at Delahaye's Champs-Élysées dealership, old Georges Morane praised François and the three drivers. He then pointed to the record-breaking car, which was displayed on a raised platform. "They used to say that our cars weren't fast. Today, this one is the fastest in the world."

While the motorsport world advanced without him on every front, Rudi continued to strengthen his leg with daily bouts of physical therapy. In late May 1934, he was scheduled to drive the W25, and he could barely cross a room without dragging his stunted leg behind him.

After Charly's funeral, Rudi holed up in his Swiss chalet, surrounded by the trinkets of his former victories. His grief left him helpless to care for himself. On frequent visits, Louis and Baby made sure that he ate and that he left the chalet once a day to go for a walk. He only went out after dusk, to avoid anybody seeing him stumbling forward on his cane. Often, when Louis was back in Paris, Baby stayed with Rudi alone. After almost withering away in this self-imposed prison, Rudi agreed to take up Louis's offer to drive the lap of honor at the Monaco Grand Prix. The race was the all-important first competition of the new formula, albeit one without any German cars.

In early April, to the applause of thousands, he drove a Mercedes convertible around the course. At the halfway point, the pain in his right leg forced him to switch to his left foot to brake and accelerate. When he returned to the harbor, the starting grid was assembled, and the engines were revving. Standing beside his car, he moved his eyes across the field. There were Chiron, Nuvolari, Varzi, Dreyfus, and twelve others. He ached to be among them and left before the second lap was finished.

"For me, there had to be a comeback . . . Otherwise life was pointless," he wrote of his thoughts when he had returned to Arosa. As a race car driver, "you are the will that controls this creature of steel; you think for it, you are in tune with its rhythm. And your brain works with the same speed and precision as this heart of steel. Or else the monster turns master over you and destroys you. I had to drive. There was nothing else for me."

While Rudi prepared for his comeback, Hans Nibel sent Brauchitsch and another driver, Ernst Henne, to test the W25. Brauchitsch mangled the car on its first outing. Henne crashed as well. Some alterations to the differential gears that controlled the separate wheel speeds solved the problem. Subsequently, Brauchitsch and Luigi Fagioli blazed tremendous times.

At 6:00 a.m. on May 24, Rudi arrived at AVUS for his trial run in the W25. He asked for the dawn start in order to avoid the press. It would be his first time behind the wheel of a race car in over a year, and he feared he might falter.

Now sheathed in a sleek white-painted aluminum body with a short tail and a headrest that tapered away from the driver, the single-seater looked fast and nimble. Rudi pulled up beside the W25 in his cabriolet, saving himself a tiring — and revealing — walk from the pits. He looked every inch his former race-car driver self, in his trademark white overalls, a scarf tied close around his neck, and his old leather racing shoes. The only difference was the walking cane he held in his hand. He could sense Neubauer and his crew inspecting his every move, no doubt eager to know if their star driver had dimmed for good.

A pair of mechanics helped Rudi into the small cockpit, causing his hip immediately to start aching. The seat had been sized to fit his body from measurements of old, but the crew had adjusted the placement of the accelerator and the brake to compensate for his now shorter right leg. As Rudi gripped the steering wheel, a company photographer snapped his profile before being quickly shooed away.

Rudi lowered his goggles and signaled with his index finger for the mechanics to start the engine. Its reverberations quickened his own heartbeat. The crew cleared back.

He shifted into first gear and eased down the track where he had claimed his first Grand Prix victory. When he came back to the straight after the north loop, he wanted to raise his fist in triumph. He was driving again.

With each lap he increased speed, the supercharger blasting at a higher and higher pitch. By the fourth, he was traveling so fast that the trees beside the track blurred. The car was extremely quick — and very powerful.

One Mercedes driver likened it to handling a fast touring car on ice. Too much throttle, too heavy a touch on the steering wheel, and it would spin around on its rear wheels or launch off the road without an instant's notice. Such was his skill that Rudi was able to handle it on his first run.

After the eighth lap, Neubauer flagged him into the pits. Before Rudi had the chance to discuss with him how the session had gone, a reporter approached the car to congratulate him. "You did 235 [kph] on the last lap," the reporter said, staring at the scribble in his notebook. "Better than they did yesterday."

"Excellent," Rudi said laconically. Nobody needed to know that he was surprised by his own performance. Nobody needed to know the pain he had suffered in achieving it.

Neubauer invited him to join the team. A requirement of doing so was joining the National Socialist Motor Corps, which Rudi understood was a paramilitary organization of the Nazi Party like the fearsome SS. He enlisted before the 1934 season began in earnest.

On July 1, the Figoni-bodied, single-seater Delahaye 138 once again circled the Montlhéry track, this time slowly and in front of 80,000 spectators in a procession of record-breaking cars. The parade was a distraction to cool the audience's heels before the real show: the French Grand Prix. Fans had made the journey from Paris in scorching temperatures to attend the mother of all races. *Motorsport* chronicled the arrival of the hordes in "decrepit old cars, sagging under the weight of humanity, [in] motor buses packed to the rear platform, with here and there a lordly Hispano or a Rolls." They joined the pedestrians on the road up to the autodrome, everybody well equipped with lunch for the long afternoon. With the Germans' silver-bodied race cars competing outside their own country for the first time, the encounter against the French and Italians promised to be epic.

The loudspeakers perched over the Montlhéry autodrome announced the arrival of the racers. René Dreyfus advanced his Bugatti T59 onto the track. The cars from Mercedes and Auto Union followed, their bodies draped with swastika flags. Over the course of the winter, René had gleaned from press reports that the Germans were pursuing victory in the upcoming 1934 season with every means at their disposal. One journalist's report in *L'Auto* described the "strange silhouette" of an Auto Union car being tested on the Nürburgring that threatened to be a "formidable competitor." This understated the truth.

The P-Wagen (P[orsche]-car) had debuted to an awed general public at the AVUS racetrack in early spring. To improve the road-handling abilities of its V16 engine, Porsche broke from convention and placed it in the rear of the car. Its driver was seated far forward on the chassis. Hans Stuck had already broken three international records with the streamlined meteor, reaching speeds of 168 mph.

Rumors were circulating about the new Mercedes. *Motorsport* repeated secondhand accounts of its remarkable speed during winter tests at the Monza track. According to spectators, "the tearing exhaust note is even more stirring than the howl of the blower gears in the famous SSKs."

As René later wrote, it was clear that the German chancellor intended "his country's cars to be supreme, the most powerful, the fastest, the most everything." René could not say the same of Ettore Bugatti. Le Patron had devoted his off-season effort toward the construction of the "automotrice," a 107-seat streamlined railcar powered by four 13-liter engines that was so fast it blew out the windows of rail stations as it passed. As for his design for the new formula, the two-seater Type 59 was a beauty with piano-wire wheels and the classic Bugatti horseshoe-shaped radiator, but it was difficult to handle on the road and it often jumped out of gear. Bugatti simply told René, "Well, it's too bad. You'll have to get used to it."

Managing the Bugatti would be even more of a challenge on the combined road and autodrome circuit at Montlhéry, which Louis Chiron noted was "scientifically designed to include every kind of corner, curve, hump and depression found on ordinary roads, as well as a fast sweep around the banking of the concrete speed oval. It was terribly bumpy, too." During practice, René had witnessed the utter superiority of the German cars on the 12.5-kilometer (7.76-mile) course. They were quicker off the line, faster down the straights, and stuck tight to the road, with their all independent wheel suspension, at speeds few had ever witnessed.

With their drivers seated far forward, the Auto Union cars looked particularly revolutionary, but both German designs were streamlined in a way that proclaimed a sudden leap into the future. Their burnished aluminum bodies, a departure from the typical German white, only accentuated this futuristic impression of cars that would soon earn the sobriquet "The Silver Arrows." René was in good company in thinking that his Bugatti resembled an antique next to them on the starting grid at Montlhéry.

The Germans were boldly vocal about their chances there. The day before the race, Manfred von Brauchitsch promised reporters, "We shall win tomorrow because we are the strongest team." By "we," the ever-confident Manfred meant himself, but the international motor press was also keenly interested in the return of Rudi Caracciola and his chances.

Whoever the winner was going to be, the press was certain that either Auto Union or Mercedes would carry away the prize for a Germany increasingly brutalized by the Nazi machine.

All the Sunday papers featured the same story. "Bloody Day of Repression" headlined *Le Matin*. The exact details were sketchy, but the presence on the streets of Berlin and Munich of heavily armed military troops "indicated serious business." According to wire reports, Ernst Röhm, leader of the brownshirts — the Nazi organization of 2.3 million members whose thugs helped propel Hitler to power — was under arrest; former chancellor General Kurt von Schleicher was dead, killed on his doorstep, apparently for resisting a warrant; and many others, maybe in the hundreds, had been arrested, hanged, shot, or committed suicide. Hitler was removing his rivals, both within the Nazi Party and without, in what was later known as "The Night of the Long Knives."

The very understandable fears the Mercedes and Auto Union teams had for their country were left unspoken. It was impossible to know who Hitler was seizing . . . or why. Nevertheless, as Manfred later wrote, such "alarming news [compared to] his desire to race made everything else appear negligible." Rudi was the same, and they had a race to run.

The engines cranked into life. The drivers could barely think over what one reporter called the "banshee wail" of the Mercedes superchargers — "the noisiest car on earth," according to *Autocar*. The Mercedes drivers stuck cotton in their ears to deaden the noise. After the starting flag fell, the band of multicolored cars howled down the low-walled straight and through the narrow opening that led from the autodrome to the road circuit.

Rudi was in first position by the westernmost bend of the course. The wheels of the Mercedes and Auto Union cars clung smoothly to the sharp turns while their competitors seemed to bump and leap across the pavement.

Nonetheless, by the time they returned to the oval bowl, Chiron was in the lead in his scarlet Alfa P3. He was greeted with cacophonous adoration by the mostly French crowd in the stands. His Ferrari teammate, Achille Varzi, was close behind, then Rudi. René stood eighth.

One car after another cracked up from the extraordinary average speeds of over 90 mph posted by the leaders. An Auto Union P-Wagen failed first, after an hour, then Brauchitsch's W25, then a Maserati. On the fourteenth lap, Luigi Fagioli's Mercedes limped into the pits and died. Two laps later, Rudi was forced to abandon his Mercedes at the far end of the course. The Mercedes team was now out of the race altogether, and the German radio commentator went silent. "The mighty German assault," Chiron later observed, "was simply melting away in the summer sunshine."

In the seventeenth lap, René's Bugatti engine began misfiring. A long pit stop appeared to solve the issue — fouled plugs — but then his car died on the next lap. Another teammate fell out soon after.

Hell-bent on winning the forty-lap race, Chiron roared around the course in his P3, throttle wide open, leaving a persistent swirl of dust behind him. Auto Union's Hans Stuck gave him a good chase, his engine's throaty vibrations almost shaking the ground beneath the car as it traveled around the ribbon of road. In lap thirty-two, he also retired because of mechanical issues, taking the last of the Silver Arrows out of competition.

The race was far from a debacle for the Germans. Anyone who saw the silver squadrons maneuver the course at seemingly impossible speeds understood that once their teething troubles were cured they might well be unbeatable.

Chiron finished first in record speed, averaging 85 mph over the course. Varzi was second, and another Ferrari driver third. It was a sweep for the Ferrari team and their Alfa Romeo P3s.

The home crowd was pleased that Louis had won, but their pleasure was bittersweet given that he was piloting an Italian car. Not a single Bugatti — the only French make in the race — had finished, nor had one even come close to the front while they were in the running. The race was a national embarrassment, spurring demands for something to be done if a French car was ever to win another Grand Prix, particularly against the Silver Arrows.

Two weeks later, under an overcast July sky, 150,000 fans descended on the Nürburgring for the German Grand Prix. The nearest town, Adenau, was too small to house the multitudes who arrived to witness the new breed of race cars that their Führer had spawned. Many had camped out in the pine-forested hillsides, and a regiment of brownshirts had marched for weeks all the way from Berlin, timing their arrival for the start of the race.

From his spot on the starting grid, Rudi Caracciola got a first-class view of a display of pageantry that rivaled any of the Nuremberg rallies. A line of swastika flags flew above the grandstand. Soldiers in jack-boots paraded back and forth by the start, and to herald the arrival of Korpsführer Hühnlein, "Supreme Chief of Nazi Motorsport," a Stuka plane flew low over the circuit. As its booming roar faded, a brass band paraded out onto the track, followed by Hühnlein, who was seated in the back of a Mercedes convertible, his path lined with NSKK motor-cycle troops wearing black crash helmets. Hühnlein was accustomed to such obeisance. A battalion commander in World War I, he had aban-doned the army to join the Nazi Party in 1920 after hearing Hitler speak of Germany "never bending, never capitulating." After the Beer Hall Putsch, Hühnlein and Hitler had spent six months together in Lands-berg prison, and they had remained close enough since then that Hühn-lein had escaped the purges the previous month.

When Hühnlein's car reached the building where the timekeepers were based for the duration of the race, he alighted, then climbed the steps to the terraced rooftop, where other Nazi high officials were al-ready waiting. Together they stood as the national anthem played. The gathered throng sang "Deutschland, Deutschland über alles, / Über alles in der Welt" as an enormous swastika flag rose over the grandstand.

When the anthem was finished, Hühnlein stepped forward in his medal-bedecked uniform to announce the start of the race. He held himself stiffly upright, shoulders bunched, almost like he was trying to gain a few inches by posture alone. At fifty-two years of age, Hühn-lein was short in stature and had a grim, purposeful face, heavy jowls, and a high tuft of blondish brown hair swept backward from his bald-ing round head. One got the impression from his look that he would rather chew stones than smile. Known to be a frank, roughneck "man of action" rather than an intellectual, Hühnlein had fulfilled the NSKK's mandate in swift order by Nazifying the national automobile clubs, starting driver schools, and recruiting hundreds of thousands of mem-bers, including Hitler Youth. He promoted the "battle for the motoriza-tion of Germany" tirelessly.

Before the start of the 1934 motorsport season, Hühnlein had sat down with both Mercedes and Auto Union and charted out their race schedules to maximize German victories. To remain in his favor, they bowed to his demands on everything from what races to run to how much affection their drivers could show their wives or girlfriends on the trackside to promoting the NSKK as "the cavalry of the future" at motor shows or political events when asked. Their drivers' celebrity was the perfect propaganda and recruitment tool. In return, the Nazi government poured money into the two automakers' coffers, financing 907,000 marks to Daimler-Benz alone, 40 percent of its team's budget since its relaunch and almost double the amount promised by Hitler.

The symbiotic relationship between the Nazi government and the two firms was on a much greater scale when it came to the rearmament of the military — a covert operation because of the terms of the Versailles Treaty. The army was increasing its ranks from 100,000 to 300,000. Air force pilots were being trained secretly by civilian aviation schools. The navy was constructing battle cruisers and submarines. The entire German economy was being steered toward providing Hitler with the tools for war. Daimler-Benz and Auto Union had already increased production of military trucks, armored vehicles, aircraft frames, and tanks. More plants were coming online. The navy and the air force wanted the automakers' engines. Daimler-Benz engineers shared advances in technology between departments. A boost in engine power or a lesson in aerodynamics in a Silver Arrow design could be used equally well in a fighter plane — and vice versa.

After the blast of a military cannon, the race began. Watching from the pits, his large signal board at the ready, Neubauer reflected on how tough a man Rudi was to manage the 172 bends of the course with his shattered leg. The Eifel race on "The Ring" required fifteen laps, the German Grand Prix a torturous twenty-five.

Although the crowds had lauded his return, Rudi had yet to notch a victory. Like his fellow Mercedes drivers, he wanted to see Germany return to its former standing in the world, but he was a contradiction to the NSKK propaganda that "motor racing is no longer primarily about personal success but a never-ending struggle for the success and honor of the Reich." Rudi joined the organization simply because it was a requirement of driving for the team, a means to an end. He wanted to win only for himself, to prove that he was back at the top of the game after his accident. All this pomp and pageantry at the German Grand Prix did not matter a whit to him.

His W25 was running faster than ever because of a new fuel the mechanics dubbed "WW." Consisting of 86 percent methyl alcohol, 8.8

percent acetone, 4.4 percent nitro-benzol, and 0.8 percent ether, it might as well have been rocket fuel.

Through the first dozen laps, Rudi held tight to the leader, Hans Stuck. In the thirteenth, he passed Stuck out in a long turn, setting a new track record. However, he had pushed the W25 too hard, and the pistons were seizing up, forcing him to retire. Stuck claimed the victory for Auto Union. Rudi was disappointed, but the race had emboldened his faith that he had lost none of his old fire.

Hühnlein cared only that it was a German car that first raced through the checkered flag, allowing him the opportunity to give another lengthy speech about the glory of the Reich before he dutifully telegrammed Hitler the great news.

More victories followed for the Germans as the Silver Arrows began to trounce the competition. During the Italian Grand Prix at Monza, Stuck again had the lead. After ten laps, Neubauer waved a flag, signaling Rudi to go faster. Already in intense pain, he cooperated nonetheless. Twenty laps. Thirty. Forty. Each time he braked, it felt like someone had jabbed a knife into his thigh. In the fifty-ninth lap, he took first position.

Stuck pulled off for a pit stop — a tire change or a refuel. Rudi wanted

MERCEDES-BENZ CLASSIC ARCHIVES

The new W25 Mercedes Silver Arrow

to go on, believing he could win now, but the strain was too much. With another fifty-seven laps to go—two more hours of driving—he pulled into the pits. He could barely mutter the words to ask Fagioli, whose own car had broken down, to take over. The mechanics needed to lift Rudi out of the car; it was impossible for him to stand.

Fagioli finished first, giving Rudi a half-share in the win, but it was little comfort. If he couldn't complete a race he would lose his spot on the team and lose his only reason to awaken each morning.

As the 1934 season approached an end, René Dreyfus was also suffering an abysmal string of performances. He had managed only one significant victory, the Belgian Grand Prix on July 29. Forty-eight hours before the race, the Germans had backed out because Belgian customs demanded a huge fee for their 3,000 liters of alcohol-based fuel, a nationalist gambit. René was stuck in third place when the two Alfas in front of him dropped out because of mechanical trouble. Therefore, his success was attributed to luck alone. The following week, in Nice, his Bugatti stalled while heading into a turn, and he crashed into a wall of straw bales. He loved being part of Bugatti, but its Type 59 was clearly outclassed against the Alfa Romeo P3s—and was in a lesser field altogether when compared to the Silver Arrows. The French press was hounding him for failing to live up to the early promise he had shown at Monaco in 1930. Although only twenty-nine, his hair was turning gray, prompting some to nickname him "the Old Man."

Journalist Georges Fraichard accused René of never having been the same since his accident at Comminges and puzzled as to whether this was because of misfortune or a lack of nerve. When René was honest with himself, he had to admit that he was missing some of his previous fire, that he was too careful. No doubt burying the charred body of his friend Stanisław Czaykowski had only further sown his fears.

René was so fed up at his string of poor finishes—and criticism— that he showed up to the starting grid of the Vichy Grand Prix still hungover from the night before. Maurice berated him for competing in such a state, and René finished a dismal fourth.

At the Swiss Grand Prix in late August, René was focused on proving to himself—and his sport—that he still deserved to be ranked among the best. Le Patron had not planned on entering the inaugural Swiss

Grand Prix, but René had convinced him, and he was the only member of the team to make the journey to Berne. "We would like René Dreyfus to be our favorite," *L'Auto* commented. "Alas, he alone will be fighting for our colors against the Italian and Germans . . . We do not think he can win in such conditions."

Every inch of the cobblestoned city of Berne seemed to be decked out for the event. Cake-shop windows were filled with race-themed confections, and the official poster, which featured a silver-and-red car in an impressionist *swoosh* before stark blue mountains, could be seen on every wall. Flags from Germany, Italy, England, and France flew from the flagpoles in front of the Bellevue Hotel, a grand nineteenth-century palace with views of the Swiss Alps.

Although everyone was staying in the same hotel, as usual — in this case, the Bellevue — René found that the other drivers rarely ventured away from their teams. The Germans stayed to themselves. The Italians and French likewise. There was no intermingling.

Practice days were the same. The Silver Arrow cars were cordoned off by rope and draped with thick cloth to avoid prying eyes. Photographers were shooed away whenever an engine hood was opened. Mechanics made sure to say nothing that might remotely give away any advantage. The German teams, especially Mercedes, brought a small army with them to the pits. Every advantage, even a half second in refuel stops or tire changes, was sought in their efforts to win — not only for the Reich but also for the government bonuses they would earn with each top finish.

On August 26 — race day — a threatening fervor took over the stands facing the 4.5-mile course that threaded through the Bremgarten forest outside the city. As had happened at the German Grand Prix, thousands wearing swastika armbands were present, and they shouted down all others.

Hans Stuck leaped into an early lead. Pushing hard, especially in the corners, René fought his way past Nuvolari and Chiron to take second position, but he could not close on Stuck's P-Wagen. Five laps from the finish, an overheated engine forced him to pull into the pits, and by the time he returned to the track, another Auto Union driver, August Momberger, had managed to pass him.

René placed third, but in the confusion over his pit stop, the pro-

French, pro-Bugatti part of the crowd thought that he had claimed second. Over the loudspeakers, the announcer stated the correct finish order. "No, Dreyfus was second," yelled his fan base. Furious at the perceived slight to Momberger, the pro-German fans went wild. Fights broke out in the grandstands. To quiet the violence, the race organizers asked René to come to the microphone to confirm that he had indeed finished third. The whole experience left him depressed. His sport had become lost in the widening chasm between countries. Races were increasingly a battleground between nations rather than individual drivers, and the Nazis were clearly investing to dominate.

After the race, Louis Chiron hinted to René that Enzo Ferrari might like him on his team the following year. A few days later, René sat down with Meo Costantini to discuss his future with Bugatti. Costantini did not try to convince him to stay. Le Patron respected René as a driver—and liked him personally—but it was time to part company.

The Bugatti team manager had some advice. He looked down at his big calloused hands, one of the few vestiges of his time as a champion driver, and spoke in his usual subdued, yet pointed, manner: "René, you could be one of the greatest drivers in the world were it not for the one thing: You are not aggressive enough. You are too steady, too dependable." His early success had come to him easily, Costantini continued, and he urged René to find something to struggle and fight for. Until that moment, greatness would elude him. The criticism cut deep, but René knew that Costantini was right.

René departed Molsheim with a watch made by Le Patron, an offer to join Scuderia Ferrari, and an announcement in the social pages that he and Chou-Chou were engaged to be married.

Two weeks before their December 8 wedding, René traveled to Châtel-Guyon to spend time with her family. They lived in the center of town in a grand villa named "La Paradou," which could easily have been mistaken for a castle with its twin steepled turrets. Chou-Chou and her father had two matters they wanted to discuss with René. First, Gilbert Miraton proposed that René consider retiring from motorsport to join his pharmaceutical firm. There were piles of money to be made—and much less risk. Perhaps in the future, René said. He wanted to continue to compete. The request had come as no surprise. Their second was. They asked that he become Catholic.

The request stung, but, on consideration, René agreed. He wanted to make Chou-Chou happy, and he knew that racing was his true religion. He refused any other definition.

Prenuptial agreements settled, the wedding preparations continued, and René traveled to Paris to buy some gifts for his fiancée. While he was away, Gilbert Miraton died suddenly, of a heart attack. He was alone at his dining-room table when he succumbed, cigarette in hand.

René rushed back to Châtel-Guyon. He and Chou-Chou decided to proceed with their wedding as scheduled, a few days after the funeral. Guests were limited to immediate family. René wore his black funeral suit; Chou-Chou, in a black dress, cried through the ceremony. They canceled the honeymoon to be with the Miraton family. It was an ill omen for their future together — at a time when the world was full of ill omens, particularly for those whose identity was cemented by the family into which they were born, whether they chose it or not.

6

The Shadow

IN MAY 1935, on a ferry to Tripoli, Libya, several drivers gathered in the bar, sharing memories and laughs. René Dreyfus was there with Tazio Nuvolari and Louis Chiron, his new Scuderia Ferrari teammates. Manfred von Brauchitsch and Luigi Fagioli from Mercedes stood across from their chief rivals of the new season, Auto Union's Hans Stuck and new team member Achille Varzi. For a brief moment, they enjoyed the camaraderie of years past.

Preferring to be alone, Rudi Caracciola snuck away. It was a starless night, and sheets of rain swept across the deck. He lingered on the aft deck beside the W25s, which were draped with gray tarpaulins. Smoking a cigarette, he thought of Charly. He had spent the winter away from Arosa, memories of her too strong in the home they had shared. He passed part of his time in the United States, where he attended the Indianapolis 500; he also lingered in Paris with Louis and Baby. By the railing, he watched the twin rivers of foam-churned sea created by the propellers. He could hear the revelry faintly from the bar and wondered who among them might die in their upcoming Grand Prix race — and what was the point of it all.

In December 1934, Berlin made clear what it felt his purpose was. From a podium in Joseph Goebbels's ministry of propaganda, Adolf Hühnlein celebrated German success in the first season of the new formula. To the crowd of drivers and Nazi officials, he declared, "Racing is and always will be the highest embodiment of motorsport and thus the highest achievement of the nation in any international competition."

Not to be outdone, other industry leaders stressed the importance of their work. The head of the national automobile club stated that the prowess of the Silver Arrows was "the benchmark for the industrial ability of a whole people." An Auto Union executive was also unequivocal, describing the engineers and workmen who had built the P-Wagen as "a community of prodigious men of German blood and soil who with tenacious will, a fierce heart, and boundless energy, labor for the kingdom of Hitler, marching proudly at its head."

These same men, Rudi knew, were moving lockstep together in the massive rearmament of Germany. In March 1935, Hitler announced to the world what most already knew: he would no longer abide by the military restrictions set out in the Versailles Treaty. He would reestablish conscription, aiming for a 500,000-strong army, and move toward the full-scale development of Germany's air force and navy. All this came at a time of ever-increasing violence within Germany.

Rudi and his fellow Mercedes drivers were certainly all aware of the atrocities committed against the "enemies of the Reich," including Jewish people, but such thoughts were usually shunted to one side. Brauchitsch recounted in his autobiography how uncomfortable he felt walking past Gestapo headquarters on Prinz-Albrecht-Strasse, knowing the rumors of the horrors committed inside. "Weren't they all communists out to destroy Germany?" he had tried to convince himself, and surely all the reports of hangings and shootings were exaggerated. Nevertheless, he decided it was best to avoid that street and in future chose to go another way. "It's none of my business," Brauchitsch wrote, with unsettling honesty. As a famous race car driver, "life was beautiful" — best not to disturb it. Rudi made the same Faustian bargain.

Rudi remained on the deck for hours it seemed, a chill curling around his spine as he thought of the risk they were taking every time they started their engines. All he knew was that he had no choice but to race, no choice but to sacrifice everything to win, and his only fear was that his body was too weak to compete against a new generation of competitors in ever faster and more powerful cars.

When they docked in Tripoli, the air was vibrating with the heat, and a cloud of red dust cast a pall over the whole city. Horse-drawn carriages took them through the narrow streets, lined with white houses and bustling with people, to the Uaddan Hotel & Casino. The star-shaped Uaddan, with its lavish gardens and fountains, was an exotic paradise compared to the staid hotels of continental Europe.

A protectorate of Italy—or its "fourth shore"—Libya was overseen by Governor Italo Balbo, a former Italian air marshal. Balbo has been described both as a "political Fascist thug" and a "hero who flew solo across the Atlantic." He loved racing, and under his theatrical touch and Mussolini's favor, the Tripoli Grand Prix became one of the richest, most extravagantly celebrated races.

The course's huge cantilevered grandstand was a showstopper in its own right, and the 8.14-mile circuit outside the city wound across salt fields, alongside the Mediterranean, past a sparkling blue lake, and around sand dunes and palm groves. A roughly shaped quadrilateral, the course allowed drivers to push flat out along the long, gentle bends. This made it very dangerous if anyone misjudged the sideways slide of their wheels toward the outer edge of the road that was inherent in turns generally taken at speeds upwards of 140 mph. Noting that this type of controlled drift required "a fine sense of balance and touch and most discriminate use of the throttle, not to mention split-second timing and a very cool head," George Monkhouse, a motorsport writer of the time, added that only a few drivers were capable of executing the maneuver.

From the very start, the 1935 race lived up to its billing as the "fastest road circuit in the world." The course, subject to the desert heat, shredded the cars' tires into angel-hair threads of rubber and caused the leaderboard to change continually as drivers were forced into the pits. First Fagioli. Then Caracciola. Varzi. Stuck.

The French and Italian factory teams were largely left behind by the Silver Arrows, which maintained average lap speeds of 120 mph. Rudi suffered three tire failures, but the Mercedes crews had reduced the average time of their pit stops, including refueling stops, from over two minutes to forty seconds—and often less. It was a military-style operation, with every move planned, coordinated, and drilled a thousand times.

Speed magazine compared their disciplined efficiency to "Kaiser Wilhelm's army of 1914." When a Mercedes driver needed to come into the pits, Neubauer waved a white flag emblazoned with a red cross. The driver killed his engine 120 yards away to keep the spark plugs from mucking up with oil. Momentum carried the silent W25 into the pits, where the driver braked on an exact mark. Any deviation from the line —an inch forward or back—cost precious seconds because the tire jacks, pressurized fuel hoses, spare tires, and starters were all prepositioned.

Then three mechanics, dressed in spotless white jumpers, launched into action. Neubauer had designated the tasks for each of them. "No. 1 gets the left rear wheel ready while No. 3 hands the driver clean goggles, a piece of chamois to clean the windscreen and a glass of water. Then he puts in the fuel. In the meantime, Mechanics No. 1 and 2 have jacked up the car and changed the rear wheels. Mechanic No. 1 has the electric start ready and the engine roars into life."

Throughout, Neubauer stood beside his driver, giving him information about the state of the race liberally mixed with praise for his elegant driving, while he also made sure that his crew filled the tank to the line and secured the new wheels with the requisite number of blows with a copper hammer.

The swiftness of the operation gave Mercedes a huge advantage over the competition—it was, as Rudi called it, "the secret of victory." Helping as well was the fact that scores of their engineers and mechanics had worked throughout the past winter to improve the power and reliability of the W25.

With five laps remaining, Varzi was leading for Auto Union, with Rudi a couple of minutes behind him in second position. Rudi knew the Italian had pushed too hard, too fast. Rudi just needed to bide his time. Two laps later, the heat vicious on their tires, Varzi threw a tread and pulled in for an emergency pit stop. By the time he returned to the track, Rudi had caught up to within a few car lengths. He patiently remained tethered there until he saw his opportunity, then sped away into the lead.

Nobody was even close as Rudi shot down the ocean side to the finish. He killed the engine and lifted his goggles from his face. Neubauer tap-danced across the pavement, then pulled Rudi free of the W25. A pair of mechanics lifted him onto their shoulders. Everyone wanted to slap him on the back or shake his hand.

He had won.

"There was the sun, the people . . . everything was good and bright and friendly, and I was back," Rudi later wrote. "Yes, that was the greatest marvel. I was back, and I could fight again as well as all the rest . . . the shadow was gone."

Reproducing the Daimler-Benz press releases almost to the letter, German newspapers headlined the victory. Their cars showed "absolute superiority" on "one of the world's toughest tracks" to "flawlessly beat" all challengers. As for "veteran" Rudi, "he has recovered and is in great shape," winning "with unbelievable bravery and perseverance" as well as "tactical precision."

A month later, in mid-June, Rudi returned to the Nürburgring for the Eifel race. Three laps from the finish, Bernd Rosemeyer from Auto Union took the lead from Rudi, right in front of the grandstands. A former motorcycle racer, the new driver on the scene was twenty-five years old, tall, blond, and handsome. With a smile that could be measured in watts, he was also cocksure, daring, and a natural behind the wheel.

This was Rosemeyer's first season at Auto Union, and only his second race as a Grand Prix driver, but he was already besting the field on a notoriously challenging course in a car that was even more notoriously difficult to handle, particularly in corners where the weight of its rear engine often led to tailspins.

Rudi stuck close to him, studying his every move. Finally he spotted a weakness in his young rival's management of the course. Near the end of the circuit, Rosemeyer always prematurely shifted into fifth gear while coming out of a swallowtail-shaped turn. In the final lap, Rudi remained in fourth gear a spell longer, then stamped on his accelerator. As they exited the turn, Rosemeyer tried to block him out, but it was no use. Fist raised, Rudi crossed the finish first.

In the pits he stood on the seat of his car, head and shoulders above the crowd that pressed around him. A sea of arms whipped upward in Sieg Heil salutes to the winner. As was expected of him, Rudi followed with his own salute, and his victory became a celebration of the Third Reich—no matter that he considered the victory his, and his alone.

At the postrace party in the Eifelerhof Hotel, Rudi watched everyone congratulate the rookie driver on his "arrival." Rosemeyer was exactly the kind of up-and-comer Rudi feared. To get in his head, the veteran champion pulled the swirl stick from his cocktail and approached Rosemeyer at his table. "Well done, my dear boy, but in the future don't be content to just drive around the circuit. Use your head." Rudi handed him the swirl stick. "With this, you can practice changing gears." Rosemeyer was stone-faced, and a bitter feud was born.

It was a darker, more ruthless Rudi who had returned to the heights of the racing world. He was more vainglorious as well. At the time, one journalist quipped that everyone needed a ladder to speak to the German champion.

Throughout the summer of 1935, René and Chou-Chou Dreyfus were on the road every week, most often in their gray Alfa Romeo cabriolet. René's wife now took on the role of timekeeper, tutored in the trade by Baby Hoffmann. The two had become fast friends, especially since Baby accompanied René's teammate Louis Chiron everywhere they went. The four made for a tight-knit band, displacing the one that Louis and Baby had shared with the Caracciolas before the 1933 Monaco accident, and then Charly's death.

The meteoric rise of the Silver Arrows had left them few victories to celebrate over their dinners together on the circuit. Not even their Scuderia Ferrari teammate "Il Maestro" Nuvolari posed much of a challenge to Mercedes and Auto Union. At Monaco, René was unable to catch Luigi Fagioli in his W25. After the race, the "Abruzzi robber" tossed the winner's bouquet of yellow roses over his shoulder like he could not be bothered with such trifles given how often they came.

Forced to be content with second place, René naively told reporters that he was hopeful for the future. At AVUS, the silver squadrons swept one-two-three, with Rudi winning again.

At the French Grand Prix, the ACF organizers added some chicanes to the Montlhéry road course to slow speeds down and give some advantage to skill over engine power. Rudi won again. Brauchitsch came in second. Due to mechanical trouble, neither Alfa P3 made it past the halfway point of the race.

For the French, it was another debacle. Not a single French make and not a single French driver even finished the race. Afterward, a newspaper cartoon depicted Mercedes mechanics performing cartwheels in the pits while a profile of Rudi Caracciola overlooked a graveyard of French cars. Charles Faroux declared the result "the great misery of French au-

Rudi wins the 1935 French Grand Prix

tomobile construction." *L'Intransigent*'s Georges Fraichard wrote about the "painful lessons" of the German victory: "When will it be understood in France that it is high time to react?" he asked.

The ACF took the coward's path, announcing its intention to run the 1936 French Grand Prix in name only. To avoid French cars being roundly beaten again, they would not allow formula models like the Silver Arrows. Instead, the event would be sports cars only. Entrants were forbidden to use superchargers and needed to be two-seaters and fitted with fenders, windscreens, horns, rearview mirrors, and a complete electric system including lights and a self-starter.

At the Belgian Grand Prix, run at the Spa-Francorchamps circuit at the edge of the Ardennes, René found himself battling against the Germans again. Try as he did to cut their lead in the turns, he could never get in front of them. Their acceleration was unmatchable. After the race, it was Rudi on top of the victory stand again, enjoying an *annus mirabilis* after his long recovery.

René spent so much time at the Silver Arrows' heels that day that he grew sick from the noxious exhaust fumes. Comparing their WW fuel to chloroform, he had to bow out at the thirty-first lap. Once in the pits, he stretched out on the pavement to gather himself. Later, eyes still burning, he bathed his face in milk and drank a glass of the same to soothe his throat. His failed effort to beat the field of German cars was not so easily cleared away.

Even with new suspension and modified engines, the P3s remained outclassed. It was only at those races that the Germans declined to attend, notably at Pau, Marne, and Dieppe, that Scuderia Ferrari had a shot. Nuvolari won at Pau, and René in the other two in tight finishes against Louis.

Between races, René and Chou-Chou stayed in northern Italy at Modena, near the small thirty-man shop that Enzo Ferrari operated from his provincial hometown. Ferrari rarely attended races, so René and the other drivers personally reported back to him on how the cars had run. Thirty-seven years old, almost six feet tall, with a formidable Roman nose, and a thick pompadour of black hair, Ferrari made an august presence. Such was his force that he rarely needed more than a soft, almost silken tone of voice to command attention.

A former Alfa Romeo driver, then one of its dealership owners, Ferrari had launched his own racing team with the Italian cars in November 1929, inspired by the Maserati brothers. He persuaded Giuseppe Campari and Tazio Nuvolari to become his first drivers. Within four years, the Scuderia Ferrari essentially took over the official Alfa Romeo factory team. Ferrari's shop, now in a two-story stone building at number 11, Viale Trento e Trieste, in Modena, brought the cars in from the factory in Milan and readied them for racing.

The new formula, and the advance of the Silver Arrows, toppled the Alfa Romeo P3 from its earlier dominance — and Scuderia Ferrari stumbled with it. Varzi left the team shortly before the 1935 season to join Auto Union, taking the spot Nuvolari wanted. The rivalry between the two was legendary. They first competed against each other as motorcycle racers. Nuvolari was raised on a farm in rural Mantua; Varzi grew up in wealthy circles in Milan. While Nuvolari was all emotion on the track, Varzi was a cold, calculating, and pitiless driver who exploited any weakness. The two could not stand to be on the same team, and Mussolini intervened to ensure that "Il Maestro" would represent Italy in an Alfa Romeo.

Such machinations were beyond René, but he was glad to be on the Scuderia. First, he and Louis Chiron were teammates. Second, he greatly

admired and liked Nuvolari, who, at forty-three, had lost none of his reflexes or his will to race. Asked by a reporter when he would retire, Nuvolari answered, "You, sir, will already be long dead before I do."

There was much to learn from him both on and off the track. Meo
Costantini had told René that he lacked aggressiveness; in that arena,
Nuvolari was the foremost professor. Away from races, René found him
modest and reserved. Besides the ever-present cigarette dangling from
his lips, he maintained a meatless diet, slept twelve hours a day, and
rarely drank alcohol — and if so, only wine. Married with two sons, he
was also a devoted family man.

Five feet, five inches tall, Nuvolari was reed-thin, with sinewy muscles drawn so tightly over his bones that he could have served as an object lesson in the anatomy of movement. Under his cropped black and
gray hair, he had the kind of face that haunted his competitors: deep
hooded eyes, a wide, long-toothed smile, a granite chin, and the inabil-
ity to mask any emotion. Not that he ever tried. Nuvolari was completely without artifice.

He also lacked a sense of caution, a fact that often saw him competing
in a plaster cast from his latest accident. During his races, he screamed,
beat the sides of his car, and rocked about the cockpit like he could will
the machine to go faster. Using his innate sense of balance and dexterity,
he slewed around corners at unparalleled speeds, virtually inventing the
four-wheel drift. "He drove like a madman, crashing often, flogging his
cars as if they were beasts of burden," one historian wrote. "He was, in
the argot of the day, the classic *garibaldino* — a driver with the slashing,
all-out style of a winner; a charger who drove with such abandon that
rumors spread through the crowds that he was haunted by a death wish
or, like Paganini, had a pact with the Devil."

At the German Grand Prix on July 28, René, who was incapacitated
with the flu, witnessed a master class from "The Flying Mantuan" in
the kind of undaunted will that it took to win. Wearing his trademark
lemon-yellow, sleeveless jersey, blue pants bound at the ankle, and a
tricolor scarf tied around his neck, Nuvolari climbed into his car. His
lucky charm — a gold tortoiseshell — hung from a chain tucked into his
shirt. Rather than a potential menace, he looked almost a dandy.

When the lights blinked from red to yellow to green, that quickly
changed. The Italian champion surged ahead on the rain-slicked Nür-

burgring course, forcing his P3 into the block of silver cars going down the initial straight. He cut the first corner like a scarlet scythe while the others veered wide. Shouting *"Corri! [Come on!],"* he shifted into third gear and accelerated. After the first lap, he was in second place, twelve seconds behind Rudi Caracciola.

Round and round The Ring, he continued his battle against Mercedes and Auto Union, their only challenger among the field of twenty. The sun came out, and the road dried. To counter the Silver Arrows' superior speed, Nuvolari nearly lived up to the legend that he never used his brakes.

Time and again, he beat them out of the blind turns only to find them closing fast on him in his side-view mirror once they were into the straights again. Up and down the ribbon of road through the pine-clad valley, Nuvolari fought, feeling like a matador against a herd of bulls.

At the midpoint, he burst ahead into the lead, but then he needed to pull into the pits. Rosemeyer, Brauchitsch, and Caracciola followed in behind him as well. With the leading four cars in the pits, the crowd anxiously waited to see who would come out first. Forty-seven seconds later, Brauchitsch returned to the circuit. Rosemeyer and Caracciola left half a minute later. Nuvolari remained.

In their excitement, the Ferrari pit crew broke the fuel pump, and they needed to pour gasoline into the tank from canisters instead. Bouncing on his feet beside his car, gesticulating wildly, cursing and scowling, Nuvolari urged them to go faster. Finally, two minutes and fourteen seconds later, he returned to the track, now in sixth position and far back.

He cut across grass berms on sharp corners and brushed straw-bale barriers to gain precious seconds. He drifted and side-slipped in the many turns, rarely tamping on his brakes. Italian journalist Giovanni Lurani described his driving as "inspired, fearless, untouchable." With three laps remaining, Nuvolari was in second, but Brauchitsch was almost a minute ahead — an insuperable gap most believed. Nuvolari chiseled eleven seconds off in the twentieth lap but only four seconds in the next. However, such was his "remorseless dance of death," as Lurani wrote, that Brauchitsch decided to forgo a desperately needed tire change. Nuvolari thrashed his Alfa about the course, closing . . . ever closing.

"The Flying Mantuan," Tazio Nuvolari, at the German Grand
Prix 1935, urging his pit crew to go faster!

The loudspeakers positioned around the course recorded the action. At the horse-shaped Karussell turn: "Brauchitsch is being closely followed by Nuvolari." Six miles from the finish, the imperious German driver threw a tire tread. Nuvolari dashed into the lead, and the announcement "Nuvolari has passed him!" stunned silent the roughly 400,000 German spectators. Minutes later, he took the checkered flag.

From his perch atop the timekeepers' stand, Korpsführer Hühnlein crumpled into a ball his prepared speech heralding another Nazi win. Such was their confidence in a German victory that the Nürburgring staff struggled to find an Italian flag to raise over the victory podium. The German newspapers chalked up the defeat to "bad luck," but everybody knew the truth. As one British editorialist wrote, Nuvolari had demonstrated "once again that the man, rather than the machine, was tempered for victory."

The crucial lesson was not lost on René, who was desperate to compete at the level he had before. Also, he had seen enough of the jingoistic show to have had his fill of it.

That fall, Lucy Schell steered her shiny new Delahaye through her hometown of Brunoy. Situated eighteen miles southeast of Paris on the former hunting grounds of French kings, the elegant town was surrounded by forests and rolling hills. For over two centuries, many of the capital's richest families bought estates in the area, their mansions monuments to their success, their noble lineage, or both. A row of the grandest stood on the rue des Vallées, which bordered the River Yerres, and Lucy and her family lived in number 26: "La Rairie."

Nobody was surprised to see Lucy driving a new car — nor were they surprised that she was driving it fast. But the two-seater convertible roadster that the residents of Brunoy saw that day was quite surprising. Straight from the Delahaye factory, it was the first of its kind: the 135.

Under the curved lines of its coachwork, designed by Joseph Figoni, the same *carrossier* who built the aerodynamic shell of the Montlhéry record-breaker, was a tiger of an engine. Jean François had further developed the design that catapulted Lucy to several wins over the past eighteen months and to third place in the 1935 Monte Carlo Rally. At 3.5 liters, the new engine was just as sturdy as its predecessor but threw out a lot more horsepower. Importantly, François had fitted it on an altered chassis, one with a longer wheelbase that stood closer to the ground. He had also improved its suspension.

Before François finished building the 135, Lucy visited Weiffenbach at his office to order a dozen for herself and friends. "There you are, Monsieur Charles. My order will cover the cost of the pair you want to run yourself. Now will you please ask François to do what he can for us?" As a mark of gratitude, Weiffenbach made sure that the very first 135 completed was delivered to Lucy Schell. Compared to some of the bone-shakers she had driven in her career, the Delahaye was a balanced and smooth ride, both in straights and in corners.

Lucy cruised through the arched entrance of La Rairie. Every detail of the three-story Belle Époque villa, from the intricate wrought-iron gate to the garden scenes in the stained-glass windows to the boathouse that looked to be carved inside a living tree, spoke to the wealth of its owners. After marrying, Lucy and Laury had purchased the house from a successful entrepreneur and made it their own, building a huge garage for their numerous cars. Their boys, Harry and Phillipe, now rambunctious teenagers, were raised there. They knew the inside of an engine better than most experienced mechanics.

As well as preparing for the 1936 Monte Carlo Rally, Lucy was already seeding plans to form a racing team to run her Delahayes. With her family money, she had the finances. Weiffenbach was finally proving to be a reliable supporter, and after years of organizing her own participation at events, she certainly had the experience to lead her own team. Wanting to chart her own course was an ambition she never lacked. Other drivers had done it over the years—Louis Chiron for one. So could she.

"Put Nuvolari in the car!" the Monza grandstands howled as Tazio Nuvolari returned on foot to the pits. Over halfway through the Italian Grand Prix on September 8, 1935, his Alfa had broken down while dueling Auto Union for first position. His countrymen were baying for him to be allowed to take over the other Alfa in the race, the one René was piloting, now in second place.

Two laps later, when René steered into the pits to refuel, Ferrari asked him to relinquish his car to Nuvolari. It was not a question, and given that Nuvolari was also the team captain, René did not hesitate to stand aside. The Italian crowd demanded that one of their own have a shot at the win, even though René had driven a marvelous race before being pulled, and many thought he could have won.

After finishing second, behind Auto Union's Hans Stuck, Nuvolari generously gave René his share of the prize money, but the race was yet another sign of how much nationalism had infected the Grand Prix. René felt that no matter how well he placed in a race, the Italian fans did not seem to be behind him. They wanted an Italian winning in an Italian car.

Worse were the sneers and openly anti-Semitic remarks he received at Monza from the crowds. The Fascists, both in Italy and in Germany, had opened the floodgates on such racial hatred.

A week after the race, Hitler announced the codification into law of such hatred during his annual Nuremberg rally. Already that year, the tide of anti-Jewish fervor had hit a high-water mark with beatings, arrests, and segregation becoming pervasive. Signs posted outside restaurants and other public places warned JEWS NOT WANTED. At Nuremberg, Hitler mandated two new policies: the Reich Citizenship Law, which stripped Jews of citizenship rights, including the right to vote; and the Law for the Protection of German Blood and German Honor, which forbade marriages and sexual relations between Jewish and non-Jewish Germans.

To top off the rally, 100,000 soldiers, an armada of planes and bombers, and numerous mechanized units, including ones with heavy artillery guns, staged a war-games show. The demonstration came at a time when Hitler was flexing his muscles on the international stage, annexing the industrial Saar territory bordering France, and securing his borders to the east.

On October 3, Mussolini further threatened peace in Europe by invading Abyssinia (modern-day Ethiopia) on the Horn of Africa. The toothless response by the League of Nations signaled that Fascist aggression might go unchecked, and half-hearted protests by France and Britain only pushed Mussolini closer to Germany. The conflict ruptured any pretense that the Grand Prix was immune from such troubles. "Politics are normally beyond the scope of this journal," the editors of *Motorsport* wrote. "But not when the sport of motor-racing is threatened. We cannot be accused of concerning ourselves with trifles, for [ours] is essentially a peacetime sport, and we presume that peace is what most people want."

Mussolini's invasion coincided with the end of the season. With several wins and top finishes in the seven Grandes Épreuves of the season, Rudi was named overall European champion. His Mercedes teammates Fagioli and Brauchitsch followed in second and third in overall points. Nuvolari was fourth. René finished fifth, with a pair of wins as well as a healthy sequence of second-, third-, and fourth-place results. As he told reporters, Tazio was an inspiration, particularly his "continual agitation" to lead every race, no matter his deficit in the field or the risk he needed to endure. In terms of prize money, René was first in class on the team.

The annual shuffle of drivers started soon after. René felt confident that he would be invited to return to the Scuderia the following season, and Ferrari confirmed that he wanted him back. René and Chou-Chou looked for an apartment to buy in Modena. Swiftly, political winds made his return impossible.

Mussolini demanded an all-Italian team, and sentiment in the country was against having a French driver in the red colors, let alone one of Jewish origin with the surname Dreyfus. It did not matter whether

René practiced the Jewish faith, or any other. As his brother Maurice, who converted to Catholicism as well, but more out of an earnest alignment with its beliefs, was told by an anti-Semite, "Your name is Dreyfus; therefore you are a Jew."

René met Ferrari at his Modena headquarters. Fumes from the machine shop hung in the air, and the screech of metal being milled made a terrible din.

There was not much to say. It was best he return to France to drive for a French team, Ferrari advised. René had no recourse but to accept this decision, much as he hated it. He and Chou-Chou packed up, sold the Alfa Romeo cabriolet, and left for Paris.

At the same time, René learned that Mercedes had signed Chiron, in part because of the recruitment efforts of Caracciola. This was a further blow to René. It was clear to everyone that his 1935 record was far superior to those of The Old Fox. Many conjectured that, given the right car, René could have been one of the top three drivers on the circuit. If anybody deserved a spot on the best team in Europe—apart from Nuvolari, who was again forced, against his wishes, by Mussolini's government to remain with Ferrari—René did.

The implication was again obvious: René was Jewish, and a Jewish driver could not represent the Mercedes or Auto Union Silver Arrows. Hitler himself had to sign off on Chiron, a non-German, joining the team. Meanwhile, over at Auto Union, Stuck was almost removed from the team because he was married to tennis star Paula von Reznicek, whose grandfather was Jewish. Only Stuck's close association with Hitler saved him from banishment. Adolf Rosenberger, the only prominent German Jewish driver, was not so fortunate, even though he had helped fund Porsche in the early development of the P-Wagen. Rosenberger was forbidden a license to compete, arrested by the SS for "racial disgrace" (code for having relations with an Aryan), and beaten in a concentration camp before he fled to America.

René had heard that Neubauer considered hiring him at one point. When it was pointed out to him that the French driver was the son of a Jew, Neubauer tartly responded, "They will decide who is Jewish." "They," meaning Hitler, sometimes overlooked a person's heritage when it suited him, but the truth was he would never allow a "Dreyfus" to race under the Nazi banner.

That October, René attended the annual ACF dinner in the club's glittering chandeliered hall overlooking the Place de la Concorde. Surrounded by the who's who of international motor racing, including a delegation from the NSKK, René endured the medal presentation to Rudi Caracciola for his "splendid" French Grand Prix win and then long-winded speeches by one elderly, tuxedo-dressed ACF official after the next. One spoke about how the League of Nations would surely guarantee peace in Europe. Another praised the Germans for their "magnificent success," avoiding any mention of their obvious government support. A third lamented the lack of a decent French car, particularly since Bugatti had said that it would probably not be competing the following year. "Are we going to resign ourselves to the decline of our colors?" the speaker asked. None spoke to how the ugly politics of Europe had cast some drivers adrift, while others benefited greatly. Seated together at the dinner, René and Rudi signaled that very divide.

To participate in the 1936 Grand Prix circuit, René knew he had few choices, if any, of teams to join. With the investments now being made to field competitive formula cars, running as an independent was impossible. He was a jockey without a horse.

With a jubilant shout of "*Partez!*" Lucy Schell launched down the palm-tree-lined road from Athens to Eleusis in her Delahaye 135. Behind her, the bleached ruins of the Acropolis, illuminated by moonlight, faded in the distance. Once out of the Greek capital, the route turned into a dirt trail that threaded past olive groves up into snow-covered mountains. Every now and then wild foxes appeared in the headlights before scattering into the darkness. It was January 25, 1936, and the Monte Carlo Rally had officially begun. Only 2,400 miles to go before Lucy and Laury reached Monaco. With their experience — and new car — they believed that victory might be theirs at last, and that Lucy would make her mark as the first female driver to win the rally.

The warm weather favored the eighteen competitors who departed from Athens. They crossed easily through several mountain passes that most years would have been deep in snow — one of the reasons why that particular route was often considered the rally's most challenging. Near Larissa, the road cut through some marshes and low fields. Traces of the Great War remained visible: ragged trench lines scarred the land,

and shell craters had left pockmarks on the terrain, which looked like a moonscape. Several times they forded streams that ran almost two feet deep.

Once into Bulgaria and Yugoslavia, the roads deteriorated into muddy ruts. To reach the next control point at the scheduled hour, Lucy had to navigate around gullies and potholes big enough to swallow their car at speeds that would have made most tremble. Their Delahaye handled these perils and the constant shifting of gears with an anthropomorphic keenness, as if it was enjoying the test of its independent front suspension and the rigor of its engine. The tires stayed tight to the road around bends, and the car practically leaped forward on the straights, its almost noiseless six-cylinder engine begging to be allowed to run free at its maximum speed of 115 mph. The cable brakes were sturdy, and the steering was as precise as a surgeon's blade.

When *Autocar* got its hands on the 135, later that year, it praised the car unreservedly: "The whole machine is responsive, almost alive, so exactly does the engine answer the driver's ideas of what he wants to do in relation to other traffic and to the actual road conditions, so exactly do the controls perform the necessary operations."

When the Schells reached Hungary, they tore down the well-kept roads toward Budapest. From there they ventured on to Vienna through a fog so thick they were unaware that night had fallen. Snow fell on the way to Salzburg, but the Delahaye kept to the road without chains.

Several days and almost a thousand miles later, they entered Monte Carlo on schedule.

That year the performance test involved a figure-of-eight course featuring two pylons that had to be rounded at flat-out speeds. The Schells posted a time of one minute, 5.4 seconds, trouncing those who had gone before. One after another of their fellow competitors failed to match their speeds, and it looked like they might have the rally in the bag when Romanian driver Petre Cristea shot around the course in his Ford V8 in one minute, five seconds flat. When his time was announced, the Schells blanched. Long-sought victory had been swept right out from under them.

They later learned that Cristea had practiced over four hundred times on a mocked-up course in Bucharest. He had also engineered his Ford to lock its wheels automatically when the steering wheel was rotated completely on either side. This facilitated his 180-degree turns at the pylons with a dexterity a journalist likened to a grasshopper's jump. Cristea's refinements stretched, but did not technically break the rules, and he and his partner Ion Zamfirescu won the fifteenth Monte Carlo Rally. The Schells had to settle for second place. Their Delahaye had performed better than they could have imagined, but they fell short.

The next month, Lucy performed dismally in the Paris–Saint-Raphaël. Her only consolation was that Germaine Rouault won. Lucy had recruited the twenty-eight-year-old Frenchwoman to drive for "Blue Buzz," her new sports-car race team.

Lucy suffered a further poor performance at the Paris–Nice Rally, another event in which she typically ranked well. In addition to competing herself, Blue Buzz was represented by Laury, Rouault, and veteran French driver Joseph Paul—all in Delahayes. Lucy was fine on the point-by-point arrivals, but the tests in between took a toll. During the 500-meter race, she mistook the pennants that indicated the finish and slowed too early. During another test, she had trouble starting her Delahaye within the allotted time. Already significantly penalized, she considered dropping out altogether. Instead, she decided to continue on to Nice, hoping to mitigate some of her disadvantage in the La Turbie hill climb.

When it was Rouault's turn, she almost vaulted up the hill. With her leopard-skin coat and pet dog in tow, she had all the panache of a speed queen, and the same ferocious competitive urge as Lucy herself.

Loudspeakers announced Rouault's impressive time just before Lucy set off. Her attempt to outdo the younger driver's time almost resulted in disaster. Lucy took a turn on the perilous course too fast and spun out of control, very nearly launching over the cliff. She finished thirty-third out of thirty-four competitors. Laury won, but that was little balm for his wife.

These defeats devastated Lucy, particularly the Monte Carlo Rally. At almost forty years of age, Lucy knew her chances of being the first woman to win it were now slim. Never one to rest idle, she sought out another goal. Early in 1936, two events, one in motorsport, the other in international affairs, brought one right into her sights.

On February 15, the AIACR, representing the major race car manufacturers, had declared the results of its winter deliberations on the new Grand Prix formula for 1937 to 1939. The previous formula, with its 750-kilogram maximum weight limit, had proved useless in controlling the speed of competitors and had only led to the dominance of the German manufacturers. Every year, their engines grew bigger and more powerful. For the 1936 season, Mercedes was fielding a supercharged, 4.7-liter engine and Auto Union a gargantuan six-liter. These overfed beasts were pumping out ratings of some 600 horsepower.

The new formula aimed to limit engine capacity to lower speeds and to allow for a broader range of car sizes, opening up participation to more manufacturers. The commission proposed a sliding scale of weight-to-engine ratios. Unsupercharged cars would have a maximum capacity of 4.5 liters matched to a minimum weight of 850 kilograms. A maximum capacity of three liters would apply to supercharged cars with the same minimum weight of 850 kilograms. A free choice of fuel was allowed.

Charles Faroux, the wise old man of French motorsport, considered the revised formula a travesty. With their limitless funds, the Germans were sure to produce supercharged engines vastly more powerful than expected. Given the amount and the type of fuel these would burn, not to mention the design complexity, such race cars would "no longer be automobiles," Faroux wrote. Further, he lamented, his country had no means to challenge the state-funded Silver Arrows, despite his best efforts in that regard.

After the debacle of the 1934 French Grand Prix, Faroux had supported the launch of the Fonds de Course, a subscription fund to aid French manufacturers in the building of a Grand Prix car. Managed by a committee assembled from the regional automobile clubs, the fund raised a kitty of 225,000 francs by selling lapel badges. This sum was split between Bugatti, Delage, and SEFAC (Société d'Étude et de Fabrication d'Automobiles de Course), the last a fledgling attempt at a national race car factory that had yet to field a working vehicle. Nothing much came of the funds, which individually were little more than what Mercedes spent on a couple of engines. Bugatti reportedly bought machine tools with its portion.

At Montlhéry in 1935, a German sweep of the podium led to the outright booing of the French government ministers in attendance. To save his country's Grand Prix hopes from the "abyss," Faroux proposed the idea of a 10-franc fee to be charged on every driving license issued. In December 1935, the government approved the measure of a "Million Franc" fund, but it would take at least a year to amass the monies, and the fight over how to distribute it had only just begun. By then, it would be of little use to finance a formula race car. When Lucy read Faroux's piece in *L'Auto* and other news articles like it, she knew something must be done much sooner.

The other event that focused Lucy's mission was Nazi Germany's occupation of the demilitarized Rhineland in March 1936. Alongside Hitler's public declaration about rearmament, this tore to shreds the Treaty of Versailles. It was an act of brinkmanship by Hitler that might well have led to a war, but the French government, already crippled by sporadic riots, an exploding budget, and a sputtering economy, chose to do nothing. A dispatch from the American embassy in Paris to Washington read: "France wants peace and fears war, does not conceal that fear, and will be forced to take the consequences." Berlin-based American correspondent William Shirer wrote in his diary, "Hitler has got away with it! France is not marching. No wonder the faces of Hitler and Göring and Blomberg and Fritsche were all smiles this noon as they sat in the royal box at the State Opera."

Through her work as a nurse in World War I, Lucy had seen what devastation a strident Germany could wreak on Europe. She had tended the wounds and miseries of innumerable French soldiers who had fallen in battle against them. There was no love lost between her and Germany, and in motor racing their Silver Arrows had wrecked any chance at victory by a French-made car — not to mention created a drumbeat of propaganda for a Nazi regime bent on supremacy.

In the wake of the new formula announcement and the Rhineland occupation, Lucy made a decision. The time had come to finish her own career as a driver and to concentrate her full attention on running her team. At that time, Blue Buzz was a limited endeavor, focused only on sports-car races and rallies, but she had loved leading its early efforts. In a bold leap, she decided to set her sights on victory in the Grand Prix. She would be the first woman to command a team in that arena, a worthy endeavor in its own right. More important, as she later told *Paris Soir,* she aimed to bolster "French prestige" and to show the Nazis that their days of unrivaled dominance were numbered. After all, nobody else was rising to challenge them in motorsport.

Taking the by now well-beaten path to the rue du Banquier, Lucy paid Weiffenbach a visit.

"What now, Madame Schell?" he asked. "Your next two cars are nearly ready; we are all soon gathering at Montlhéry for a public demonstration . . . What more can Delahaye do for you?"

"Very simple, Monsieur Charles!" Lucy said. "You can build me a 4.5-liter racing car for the 1937 formula. I've decided to run a team in the Grand Prix."

Weiffenbach was speechless.

She told him the new name she had chosen for the team, Écurie Bleue (Team Blue). It was pure call-out to the color that traditionally represented France, dating back to the early days of the monarchy. "They will be my cars," Lucy continued. "I will finance the project from top to bottom: design, construction, development, the racing itself. I'm offering you an opportunity which I'm sure isn't available to any other firm in France. Now, what do you say?"

7
A Very Good Story

WHEN LUCY SCHELL left the office of Charles Weiffenbach that day, there was much to consider. If anybody else had asked him to build a Grand Prix car, he would have laughed them out of his office. He was a businessman, and when Madame Desmarais instructed him to build cars that won races, she meant sports cars that spectators might imagine taking out on the open road on weekends. Grand Prix cars were too powerful and temperamental for the typical driver to handle. Further, they were very expensive to design and produce, let alone maintain over a season. Firms like Daimler-Benz and Auto Union could afford such investments. They were massive industrial complexes with numerous plants, legions of engineers, vast sums of money, and clear government support against which Delahaye was a Lilliputian.

But Weiffenbach had come to expect these demands from Lucy, and he recognized an opportunity when it crossed his path. As a young man in his twenties, he was about to leave for a job in Indochina when he had run into a friend who was considering applying to be Delahaye's chief of production. Although Weiffenbach had limited technical education and had spent only three years at Ravasse, an engineering firm that had worked on the Léon Bollée tricycle cars, he threw his hat into the ring. He won the position, and his friend took the job overseas.

On the surface, Lucy had presented an outlandish idea, but with her financial support the development of a Grand Prix design would not cost his company a single franc. If they managed to produce a worthy car, the publicity alone would prove a windfall unlike any since the board charged him with making "the marque better known." Most of all, Weiffenbach liked the idea of returning France to its rightful pedestal atop motorsport. The patriotic name Lucy had chosen for her team sealed the deal for him.

He tasked Jean François to begin development. They both knew that a supercharged three-liter engine, particularly one built by experienced German engineers, was sure to pump out far more horsepower than a 4.5-liter unblown engine. The formula's authors thought they were equalizing the advantage of superchargers by the greater capacity of their unblown counterpart, but their ratios were surely off.

Designing a supercharged engine was a very costly, time-consuming effort in which they had no experience. Lucy Schell might have deep pockets, but not that deep. Furthermore, Weiffenbach may have been excited over the idea of competing on the Grand Prix circuit, but that did not mean he was suddenly seeing the world through rose-tinted glasses. Any design worth doing must serve as the basis for a car that the general public would buy, and a blown engine would be of little use in a regular Delahaye production model, not least because of its high fuel consumption.

François proposed the idea of a V12, unsupercharged engine sized to the formula's maximum capacity of 4.5 liters. He had already done some work on a design for a sports car, and this could be adapted for the Grand Prix model. Weiffenbach added that it had to be lightweight, efficient, durable, easy to manufacture or fix, and fueled by regular gas.

Over the next few days, François had long working lunches at the Restaurant Duplantin, sketching designs in his notebook. The bustle of life on the Place Péreire outside might as well have disappeared as he worked out his rough plans in pencil. Sometimes he became so excited that he scribbled on the white tablecloth. They were only drawings — and only of the engine. The chassis, suspension, brakes, steering, transmission, aerodynamic body, and a thousand other considerations had to be decided. But it was a purposeful start.

In the spring of 1936, René Dreyfus was working for a Delahaye competitor, the Talbot-Lago automobile company. He spent most of his time in its well-appointed offices in Suresnes, a suburb west of Paris. The sign on the door to his office read, in elegant script, CHEF D'ÉQUIPE (team leader). While René's racing brethren were preparing for the new season, he was wearing a suit and tie and working on creating a factory team for the fledgling manufacturer to compete in sports-car races.

Whenever anyone asked, René was positive about their prospects: he only needed to hire one more driver; their new car was handling well; their schedule of events was all set. "We still have a lot to do," he declared brightly to a reporter, but they would be ready for their first challenge in early May.

The truth was that he had been forced to hire and babysit an inexperienced driver named Jimmy Bradley, solely because he was the son of the *Autocar* editor. Worse, and much more troublesome, their car remained hampered by mechanical trouble.

The previous fall, with no other options, René had agreed to become the Talbot race team captain. Company president Anthony Lago almost made it seem like a good idea. Lago was a garrulous, handsome Italian in his early forties, and a first-class charmer in several languages. One of the fifty founding members of Mussolini's National Fascist Party, Lago stood up to Il Duce when he mutated from being a people's fighter into a militant bent on dictatorship. Lago escaped an assassination attempt by blackshirts, tossing a hand grenade at his would-be killers. He fled Italy. In 1923, after a brief period studying engineering in Paris, Lago emigrated to England. Within ten years, driven by a kinetic energy, he built himself up to be the kind of businessman whom boards bet their companies on.

Lago intended Talbot to be his *pièce de résistance*. Formed in 1903, the company dominated motor racing in its early years, earning its cars the moniker "The Invincible Talbots." After World War I, a muddle of reorganizations and mergers had left the original company a mess. Factories in France and England both produced vehicles, often at cross-purposes. Its name, Sunbeam-Talbot-Darracq (STD), reflected the lack of focus that propelled the company into bankruptcy by 1934. Its new owners were prepared to sell off the French part of the business when Lago secured a deal to revitalize the Suresnes plant for a modest salary and a share of the profits. By the end of 1935, he owned the company's continental arm outright and added his surname to the marque. Such was his deft deal-making that he had spent none of his own money making it happen.

As for his strategy to resuscitate the company, Lago was clear. He would cut expenses, produce lighter, more attractive cars, and, as he declared to *L'Auto,* "in the very near future, we will resume the place Talbot has held so brilliant for a long time in international events." He tapped René, a premier Grand Prix driver who had not been recruited to any team, to lead their efforts and ordered his engineers to ready a four-liter version of their six-cylinder T150 for the 1936 sports-car season. This would include the French Grand Prix, whose organizers had abandoned the international formula again to prevent a Silver Arrows sweep.

Many rightly argued that it would be a Grand Prix in name only, but no different from the other French events that year that would run only sports cars, including Marne and Comminges. Only Pau, a race garnering more attention each year, would be run under the 750-kilogram formula, but with the Germans fielding no cars, it was severely diminished. In addition, Mussolini forbade any Italian teams from participating in France as long as the League of Nations continued to tighten sanctions against his country in response to his Abyssinian invasion. Yet again, politics trumped sport.

On May 24, René changed his business suit for overalls to compete in his first competition for Talbot: the Three Hours of Marseille. His blue T150 had been fitted with mudguards and lights to match the sports-car specifications. A reporter likened its flat and utilitarian design to a pontoon boat.

At the Miramas autodrome, René and his fellow Talbot driver André Morel faced ten Delahaye 135s and a single Bugatti to see who could cover the greatest distance around the track in three hours. Two of the Delahayes were fielded by its factory team, and three by Lucy Schell's Blue Buzz, including one piloted by Laury. Before the race started, Tony Lago pulled René aside. "Just go as fast as you can," he said. "It's okay if you break down." They both knew that their cars were unreliable, and all Lago wanted to achieve from the race was to prove that Talbot was back in the game.

René followed the orders he had been given — and then some. After the first round of the five-kilometer circuit, he was leading by 300 meters. He clocked a lap record, hitting speeds of over 125 mph. After the fifth lap, he retired with engine trouble. Morel suffered the same fate and bowed out too. Delahaye dominated, its factory driver Michel Paris (the racing name for Henri Toulouse) winning the race; Laury Schell placed second.

Two days after the race, a strike broke out in an aircraft factory outside Paris. From there, walkouts and sit-ins spread across the city and then enveloped the whole country. The economy ground to a halt, and the value of the franc nosedived. Revolution was in the air, and the red tide of communism looked like it might prevail. No corner of France was immune. Lago was forced to juggle a host of creditors to save Talbot. In Paris alone, 350,000 workers abandoned their jobs to march in the streets. Cafés and department stores shuttered. Hairdressers laid down their scissors, shoe-shiners their brushes.

Ettore Bugatti had long considered his staff to be like his children. "I've nothing to worry about," he remarked after the strikes hit. Soon, Molsheim was in revolt, and his workers took over the factory, refusing him entry through his own doors. Crushed, Bugatti departed for Paris, leaving Jean, his more than competent twenty-six-year-old son, in charge.

Because of the strikes, the 24 Hours of Le Mans was canceled for the first time in its history. Somehow, the ACF managed to stage the French Grand Prix, at Montlhéry on June 28, albeit before a meager crowd. (A horse race that same day sold more tickets.)

René led a team of three Talbots, there were nine Delahaye 135s, and Jean Bugatti showed up with a new sports car of his own design. Wider and longer than the Talbot or the Delahaye, its magnesium-alloy body so resembled a streamlined tank that this was what Bugatti decided to call it: the Type 57G Tank. The veteran French champion Robert Benoist and Jean-Pierre Wimille each piloted one.

"Your job will be to stay ahead of the Bugattis for as long as you can," Lago had instructed René. "That's all I want."

At ten o'clock, thirty-seven sports cars lined up diagonally by the start, their drivers standing a short distance away, upper bodies bent forward, ready to dash to their vehicles for the Le Mans–style launch of the race. At the pop of the gun, René hurried to his car. He jumped inside the cockpit, started the engine, and was away first. As he had done at

Marseille, he finished the initial lap in the lead, but the race was 1,000 kilometers long and would take eight hours to finish—if they were quick.

René fell behind Benoist in the second lap, then was overtaken by Wimille, but he managed to hound the Bugattis for a fair distance. He also set a lap record. Before the halfway mark, the three Talbots each fell victim to a litany of minor mechanical troubles—plugged fuel lines, faulty spark plugs, displaced push rods—that added up to crippling delays.

In the end, Wimille won for Bugatti, and Michel Paris came in second for Delahaye. Talbot finished eighth, ninth, and tenth. Lago was thrilled at the performance. His team captain had again proven the T150's unmatched speed. René felt differently. He wanted to win—and he wanted to win on the stage of a true Grand Prix.

At the Marne race on July 5, René again outpaced his competitors in the early stages, posting his third lap record in as many races on the twenty-third round of the 4.8-mile road circuit at Reims-Gueux. Lago celebrated on the sidelines, believing that victory was in hand. Two laps later, René drifted into the grass a couple of hundred yards short of the pits. His crankshaft had broken, and his engine was dead. Wimille won again in the Tank.

The Delahaye factory team suffered two crashes at that race. Paris, their best driver, broke his back after hurtling off the road at a sharp corner in the thirty-first lap. Rushed to a clinic, he survived but was paralyzed. A few laps later, his teammate Albert Perrot slid off the road into a ditch. His 135 was crushed—he emerged bruised and dazed but alive. Delahaye had only narrowly avoided losing both its drivers.

At the dinner afterward, René asked Wimille if Bugatti was planning to race at Comminges the following month. He hoped that by then the Talbots would be free of mechanical trouble. "Why, yes, I believe so," his cocksure rival said. "We'll be there to see you off again!"

Ever since they shared a hospital room after their accidents at the 1932 Comminges, René had cared little for the lanky, taciturn driver, in part because of a conversation they had about their respective futures. At the time, René had no plans other than to race cars. Wimille wanted to enter politics. When René asked what platform he would represent—and who would vote for him—Wimille simply answered, "Women." He was that sure of his handsome, rugged looks and charm. The son of a famous and well-heeled French racing journalist who wrote for the *New York Herald Tribune,* Wimille had an easy entry into the sport. Humility was lost on him.

On August 9, they met again at Comminges. In the eleventh lap, René was just readying to overtake the lead from Bugatti when his left front stub axle suddenly snapped. The wheel shot away, and his Talbot careered sideways across the road, skittered over a grass embankment, and came to a grating halt against the side of a house. René emerged from the cockpit unharmed but shaken. Time and experience had made him less prone to believe in the workings of chance, but he felt certain that he needed to leave Talbot to improve his luck and leave the crash behind him.

He finished the season without a single win and enraged over being little more than an outcast on the Grand Prix. Had he been on the official Ferrari team, he would soon be heading off for the Vanderbilt Cup in New York, the revival of a once-famous American race. Instead, he was stuck in Paris, where strikes and unease over Germany persisted. He feared for the future, much like all of France.

While René was competing in French sports-car races, the Grand Prix drew into a contest between the two German automakers. On May 10, 1936, Rudi Caracciola was in distant fourth as the drivers approached the end of the Tripoli Grand Prix. This was his second race of the season. He had won at Monaco, but here, on the fast Mellaha circuit, there was no catching Hans Stuck, who was in the lead. Only Achille Varzi, also in an Auto Union P-Wagen, was even close. With their six-liter engines and improved handling, their cars were unstoppable.

Watching with Governor-General Italo Balbo in his raised booth were two officials from Hitler's inner circle, Philipp Bouhler and Martin Bormann, their presence evidence of how tight the German-Italian axis had become since the Abyssinian invasion.

As Stuck passed the pits, his team manager, Karl Feuereissen, waved a green flag: the signal to slow down. Given Stuck's lead, it was a wise strategy; there was no sense in pushing too fast and risking an accident. To Stuck's shock, Varzi soon closed on him. As they were teammates,

Varzi should have been given the same signal to ease back. Moments later, he overtook Stuck and he did not look back. In the final lap, he posted a new course record.

Stuck finished 4.4 seconds later. As soon as he reached the pits, he leaped from his car, removable steering wheel in hand, and yelled about how that "bloody swine" Varzi had stolen his race. An Auto Union mechanic tried to calm him, telling him that Varzi was never given the signal to slow down. In fact, he had been flashed the red flag—to go faster. Now at a boil, Stuck confronted his team manager, demanding an explanation.

Feuereissen drew him aside and quietly said, "I had strict instructions."

Hans was confused: What instructions? From whom?

"Berlin and Rome have decided that, wherever possible, Italians should win Italian events, even if they're driving a German car." Hans threw the steering wheel at his manager's feet before storming off.

Rumors of their conversation reached Neubauer, then Rudi and the rest of the Mercedes team. That evening, Governor Balbo was holding his annual postrace party, and there were sure to be fireworks.

Rudi arrived with Louis Chiron and Baby. They were almost inseparable again, now that the Monégasque driver had joined Mercedes, a recruitment that Rudi had orchestrated largely to be able to spend more time with his friend's girlfriend. Although Louis was oblivious to the fact, Rudi had developed feelings for Baby. She was always full of energy, and her friendship had seen him through his worst days. Early that spring, in Paris, Rudi had tried to confess his love for her: "If I could find a girl like you somewhere, I'd marry again." Baby countered that he would do better to find a "nice young girl."

Arriving at the Tripoli palace, the three found a scene straight from *One Thousand and One Nights.* Local soldiers dressed in rainbow-colored uniforms and carrying ceremonial scimitars stood guard on white horses at either side of the wrought-iron gate. More of them lined the wide stairway leading into the palace. Everywhere, as Alfred Neubauer described the scene, were "chalk-white Moorish façades, crenellated walls, pointed gateways and soaring pillars that stood out like filigree against the starry sky." A band played on a marble terrace above the gardens and fountains, which were all aglitter with light. In a long reflecting pool, half-naked dancers pirouetted in the water.

Dinner was served in a large hall, where over 150 guests settled around a horseshoe-shaped table heaped with platters of food and drink. Governor Balbo sat at the center of the curve, with Hans Stuck to his right—a place of honor that should rightly have been set for Varzi. After hors d'oeuvres, Balbo stood and held his champagne glass aloft. Looking at Stuck, he bellowed, "A toast to the victor of the day."

Calmer now than before, Stuck tried to protest: Varzi had after all won the race. Balbo would have none of it. "You can't fool me, Herr Stuck. I saw very clearly how you were pulled back so that Varzi could go into the lead. I don't like little deals like that. Don't like them at all. Politics should be kept out of sport, and I emphasize once again that I consider you the true winner of this Grand Prix."

Several seats away, a glass broke, and Varzi stormed from the hall. That evening he would take morphine for the first time, a drug that almost ruined his life.

Although embarrassed by the scene, Rudi now accepted that politics was inseparable from racing. He remained ambiguous about the Nazis, and on a warning from a friend that Hitler was intent on war, he decided to move permanently to neutral Switzerland. Still, if he wanted to compete, he needed to toe the Nazi line and play the role of one of its heroes. Whenever Rudi mounted the victory stand, he vigorously sang "Deutschland, Deutschland Über Alles," and he was content to appear in Reich propaganda that labeled its Grand Prix drivers "swift as greyhounds, tough as leather, strong as Krupp steel."

The next weekend, in Tunis, Rudi won his second race of the season, but his reign as European Champion—and Germany's champion—was about to be challenged in earnest by Bernd Rosemeyer. Their duels would set the stage for the lengths to which Mercedes and Auto Union would go to shine brighter in Hitler's eyes and rule every corner of motorsport.

A blanket of fog settled over The Ring during the seventh lap of the Eifel race, on June 14, and the Nürburg castle disappeared in white cloud. The mist spread across the grandstands, and the scoreboard and signal system faded away. Visibility on the roller-coaster mountain course dropped to fifty yards. Bernd Rosemeyer maintained his speed through the blind twists, distancing himself from his closest competitor, Nuvolari in a red Alfa.

The crowds lining the course heard the thunder of the Auto Union V16 engine. Then, all of a sudden, the silver car broke through the wall of fog only to vanish just as quickly. "It must be a drive of amazing peril, groping through the clouds, in the mountains," *Autocar* reported. At the finish, the writer continued, "no one can see the approach of the cars. The staccato bark of the Auto Union is heard at last, and the crowd cheers Rosemeyer to the echo."

The win was the young competitor's first of the 1936 season, and his driving mesmerized all who witnessed it. Racing journalists called him "Der Nebelmeister" (The Fog Master), echoing the nickname "Der Regenmeister" that Rudi had earned at his rainy triumph at AVUS a decade before. Afterwards, Bernd's jovial smile beamed from the newsstands, and profiles of the "thunderbolt known as Rosemeyer" spun from the presses.

Bernd had been watching automobiles get disassembled in his father's garage since the time he could crawl. At the age of nine, he was so eager

Bernd Rosemeyer on "The Ring" in his rear-engine P-Wagen, 1936

to drive that his parents fixed wood blocks to the pedals of a car and let him have a go. By sixteen, he had saved enough pocket money to buy his first motorcycle, a 200-cc DKW. His persistent speeding around the town caused the police to revoke his license. "Just as a bird needs the air, and a fish needs the water, so Bernd Rosemeyer needs his motorbike," he once wrote.

In the spring of 1934, Auto Union recruited him to its motorcycle team. Many victories followed. That fall, they invited him to try out for the race car team. Depending on the account, he either wore his best Sunday suit or a new set of racing leathers to the trials—Bernd always liked to stand out. Despite having never driven a Grand Prix car, he finished with the second-fastest time and earned a spot on the team.

With his fast reflexes, ferocious driving style, and natural balance behind the wheel, Bernd reminded many of Tazio Nuvolari. During one race, a stuck brake sent him off the road. He avoided crashing by threading his car between a house and a telegraph pole. There was only a sliver of clearance on either side, but Bernd managed it: "Like a high-speed camel going through the eye of a needle," one writer observed.

Bernd won his first Grand Prix race in the last event of his rookie season, the 1935 Czechoslovakian Grand Prix, run at the Masaryk circuit outside Brno on September 29. At the celebration party afterward, he met and fell in love with aviator Elly Beinhorn. Already famous, the twenty-eight-year-old was giving a talk in the Czech city about her latest aerial expedition. Always in a rush, Bernd asked her to marry him soon after meeting her. The newspapers labeled them "the fastest couple in the world." In July 1936, a month after he won at the Nürburgring, Bernd and Elly wed.

Bernd Rosemeyer was the living embodiment of the Nazi ideal. As historian Anthony Pritchard remarked, "If [he] had not existed, then the National Socialist party would have had to invent him." While he was racing motorcycles, he had worn a swastika armband, and like many ambitious young men, he had joined the SS. As a reward for his Eifel race victory, Heinrich Himmler personally promoted him to Obersturm-führer (senior leader).

Caracciola continued to be Germany's best driver, but in comparison he was a crippled old veteran. Bernd was, the newspapers spun, "the radiant boy," a "bold fighter" as well as "a man of action," "risky but self-assured." His looks, the Aryan ideal from head to foot, did not go unremarked: "Beautiful blond Bernd" was a charming rascal, one report went. "Unforgettable, dazzling Bernd, the young Siegfried among the racing aces in the world," gushed another.

With his wife, Elly, the convention-breaking "German heroine" who had survived a crash landing in the Sahara Desert and who had flown solo from Europe to Australia, they were a fairy tale of the Reich, proof of the superiority of "Aryan blood." The dashing race car driver and adventurous aviator were seen as "Das Traumpaar" (the perfect couple), and together they captured the popular imagination like few in the Reich had done before.

Thirteen days after Bernd and Elly's marriage, Rudi faced his rival at the German Grand Prix. Bernd took a significant lead by the first half of the race and was never challenged. He spent his final lap waving at his adoring fans and finished four minutes ahead of his closest competitor. In front of photographers, he sealed his commanding triumph with a kiss from Elly. An exuberant Hühnlein presented him with the trophy and a laurel wreath.

Throughout the award ceremony, Rudi stood grimly beside Bernd. The Summer Olympic Games in Berlin would put their mounting rivalry on hold, and what was more, it would show the world how powerful a force sport had become in relations between nations.

On August 1, Adolf Hitler arrived at the Olympiastadion, a concrete coliseum festooned with swastika banners, to the peal of thirty trumpets. The *Hindenburg* airship floated above. As he crossed the stadium, many of the 110,000 spectators stretched out their arms and shouted "Sieg Heil" to a thunderous cadence.

A young girl in a white dress greeted him halfway across the field. She handed him a bouquet of flowers and then kneeled in obeisance. Hitler then mounted the steps to his high balcony while composer Richard Strauss led a choir of 3,000 singing his "Olympic Hymn."

Once Hitler was seated, groups of athletes from forty-nine nations entered the stadium. They paraded before the German leader, arms raised in an Olympic salute that unnervingly resembled the Nazi raised arm. Finally, Hitler stood before a microphone and declared the Games open. Twenty thousand doves were released — ostensibly a symbol of peace among nations — and fluttered about. Those citizens caught in the recent outbreak of the Spanish Civil War would have seen their promise as a false one.

High above the stadium, the Olympic flag unfurled. Then a bell, deep and resonant, tolled, followed by a blast of cannons that sent the doves into a whirling panic. A slight German runner in a white singlet and shorts circled the stadium with the Olympic torch blazing in his hand, then dashed up the steps beside the Marathon Gate to a tripod basin to which he touched the torch. Billowing flames rose up from the basin.

Over the next two weeks, the Games enthralled Berlin — and much of the world. Millions listened to the events on radio broadcasts or watched on cinema newsreels. Each night, Nazi officials threw extravagant parties to entertain diplomats, journalists, athletes, celebrities, business tycoons, and other dignitaries. "Champagne flowed like water," wrote one guest. Reich aviation minister Hermann Göring hosted an evening that included stunt airplane flights and Renaissance costumes.

Throughout the Games, Hitler was omnipresent. "Face contorted, he observed the performances of his athletes with a passionate interest," the French ambassador remarked. "When they won, he beamed, slapping his thighs loudly and laughing as he looked at Goebbels; when they lost, his expression hardened and he scowled." Given the strong state support of its athletes, the Germans dominated the point rankings. *Der Angriff,* the Nazi party rag, crowed that it was "truly difficult to endure so much joy" over the German medal count.

The Nazis presented a "New Germany" minus its worst militaristic and racial instincts. They allowed soldiers to wear civilian clothes, stowed away signs stating JEWS AND ANIMALS NOT ALLOWED, placed on hold vitriolic attacks against non-Aryans in newspapers, and even returned to library shelves some books authored by Jews that had been burned in earlier purges. Despite the Nazis allowing a single Jew, fencer Helene Mayer, to join the Olympic team — and this a token gesture conceded only after a boycott — Victor Klemperer, a Jewish professor living in Dresden, noted in his diary that foreigners came away with the impression that they were "witnessing the revival, the flowering, the new spirit, the unity, the steadfastness and magnificence, pacific too, of course, spirit of the Third Reich, which lovingly embraces the whole world." Klemperer, however, was not fooled. In the same entry, he wrote, "I find the Olympics so odious because they are not about sports—in this country I mean—but are an entirely political enterprise." This was an understatement. Time and again, Nazi propaganda called athletes "warriors for Germany" who stood "at the front lines of foreign policy."

After leaving Berlin, a *L'Auto* columnist wrote about the "great lesson" of the Games: athletic success was now indelibly intertwined with the prestige and power of a nation. "The Age of Sport is now consecrated; deplore it or approve it, but you will not change it." And within the sporting arena in 1936, the Germans looked unbeatable—from the Berlin Olympics to Max Schmeling's twelfth-round knockout of Joe Louis in Yankee Stadium to the Silver Arrows' dominance on racetracks across Europe.

To design any car, let alone a Grand Prix competitor to combat Mercedes and Auto Union, demanded many things: the inspirational leaps of an artist and the deductive patience of a mechanic, as well as insights into metallurgy, electricity, physics, mathematics, aerodynamics, production, and, of course, engineering.

There were the geniuses, like Ettore Bugatti, who relied on an intuitive feel for what worked. But for most, innate vision needed to be buttressed with academic study and long apprenticeship, all to understand what automotive historian L.J.K. Setright called the essential truth of the business: "A car is not a thing, it is an aggregation of things, a compound complex of numerous, mutually supporting components that are infuriating because they are also mutually interfering. The man who can see how to eliminate these incompatibilities, how to make each component in such a way that it does its various tasks as well as can be while detracting from the performance of all the other components as little as can be, can see how to design a car."

Instinct, study, experience—all three influenced Jean François in his approach to what Delahaye was calling its Grand Prix car: the 145. A strong dose of his own practicality was also added to the brew.

Everything started with the engine. From his first sketches, made at Restaurant Duplantin, a V12 was the obvious choice, even though the company had not built one since its boat-racing days. Henry Ford figured that an engine shouldn't have "any more cylinders than a cow has teats," but he was in the minority. Most valued the V12 design. Although complex to build, it generally ran more smoothly, allowed for higher revs, and delivered more power than a six-cylinder engine of similar capacity. A driver once described the V12 as delivering "a peculiar pulse that is the sonic equivalent of strawberry mousse and cream." There was simply something smooth and luscious about it.

Using the proven method, François set the two six-cylinder banks at a 60-degree angle to one another. Each cylinder had a bore of 75 millimeters, a measurement that matched the famous French field guns of World War I ("Les Soixante-Quinze," the "75s"). With a maximum capacity of 4.5 liters, this left a low piston stroke in each cylinder of 84.7 millimeters, almost one-quarter of the length shorter than the Delahaye 135 engine.

The two engines were very different beasts, however, and François could not base the Delahaye 145 on an old truck engine. He needed to venture into new territory. One of his first creative leaps came in the design of the camshaft. His V12 would actually have three: one in the center to operate the intake valves, feeding fuel and air; and the other two placed on the sides of each cylinder bank to operate the exhaust valves from the engine. This would allow the engine to breathe better and operate more efficiently. Its crankshaft was similarly engineered to maximize performance. In another original move, François designed the one-piece cylinder block to be cast in magnesium alloy instead of traditional aluminum, reducing its weight by 35 percent.

Twin magnetos; twenty-four spark plugs; water, oil, and fuel pumps; a trio of carburetors; timing gears; roller bearings—these and dozens of other parts were also needed to create the engine. François's early simple sketches turned into scores of iterative blueprints, enough to architect a city skyscraper, until he had finished plans ready to send to the factory floor.

François knew that his 4.5-liter design, regardless of how well built and tuned, would produce a maximum of 225–250 horsepower, a measurement almost half of what the Germans would likely produce from a three-liter, supercharged engine, particularly one operating on a custom fuel mix. But this was the new formula, and Monsieur Charles wanted an engine that could be installed in everything from a Grand Prix competitor to a sports car to a high-end coupé used for outings about the city or countryside. That was the Delahaye way.

Like their "son-of-a-truck" engine that had proved so successful when breaking speed and endurance records, the 145 would have efficiency, versatility, and toughness on its side. In a long race, these qualities were to be prized. Such an engine was useless without a chassis that could nimbly handle the road. In its design, François mirrored closely that of the 135 — its innovative leaps had resulted in a command of everything from oval tracks to mountain climbs. A two-seater of rigid ladder-frame construction, the chassis sat a bit lower and ran slightly longer than its predecessor. But these were minor differences. They shared a similar independent front suspension, cable-operated drum brakes, and a retracted engine position to distribute the car's weight evenly. "No wild innovations here," one critic would write. "Just well-polished state-of-art logic as a foundation for refined tuning, development, and durability." A small team of draftsmen and engineers created another shelf's worth of blueprints, each several feet wide and many feet long when unfurled. François would have to hire a coach-builder since Delahaye lacked an in-house operation, but the car's body would be lightweight and utilitarian.

The factory had four 145s on order from Écurie Bleue. Given the strikes and the widespread labor trouble, building them had been slow, but the first chassis was on course to be assembled by late fall 1936, and the company expected to have an engine ready for testing in early 1937. The finished cars, weighing roughly 850 kilograms to meet the formula minimum, were to be ready by late spring. Proof of whether the design was successful would emerge on the Montlhéry track.

Before that time, Lucy Schell needed to find a worthy driver.

That September, Lucy finished up her first season running her own team with the Royal Automobile Club's 1936 Tourist Trophy in Belfast, Northern Ireland. Her two Delahaye 135s had failed to claim a top position, but they had competed in their maiden race outside France — another mark in the ledger of their progress.

Since spinning out of control on the La Turbie hill climb at the end of the Paris–Nice Rally, ruining her chances of a win, Lucy had followed through on her intention to devote herself completely to running her team. Her role included handling the drivers, managing the partnership with Delahaye, building out an organization of mechanics and support staff, and bankrolling the whole affair.

There was much to show for her efforts. Her two best drivers — her husband Laury and Joseph Paul — ranked well at the French Grand Prix and at several other sports-car events in France. The team now had its own garage, a two-car transport truck, and a mobile machine shop. Weiffenbach had decided not to have a factory team after the two crashes at Marne; as a consequence, in an arrangement similar to Enzo Ferrari's with Alfa Romeo, Lucy's was the only team racing Delahayes.

After returning to France from Northern Ireland, Lucy began preparing for the 1937 season. She planned to field several Delahaye 135s for rallies and sports-car races. For the Grand Prix, the implementation of the new formula had been delayed until 1938 because of quibbling over final details. Since Jean François had much to do to turn his sheaf of blueprints into a working race car, this suited Lucy fine. Nonetheless, she wanted to start the search for and recruitment of a Grand Prix champion to lead her team.

Such drivers were few in number, and fewer still were those not already committed to established teams. Lucy was an inspiring force, but even if a Nuvolari or a Chiron could stomach taking orders from a woman, they would never take the leap to drive for an untested, independent operation whose cars were in development. She needed someone desperate, someone eager to prove or reestablish himself.

One name, and one name alone, came to mind: René Dreyfus. Anyone could see that he was not happy at Talbot-Lago. There was no place for him at Bugatti. Neither the Italians nor the Germans would have him. Since she formed her sports-car team, Lucy and René had crossed paths more often than in the past, though they rarely shared more than the short conversation they had at the victor's table after the Paris–Nice Rally in 1934. Nonetheless, she had seen him race dozens of times. He was a driver, most of the press agreed, with "finesse and intelligence," as well as one of "great precision who could teach a class on holding a line." As a competitor, he was "calm, measured of movement, and patient as an angel." Lucy knew he had the skill to win any race — if he had the fire anymore to do it — and she was nothing if not a fire-starter.

Lucy liked to conduct her business at her home in Brunoy. She would usually invite a prospect to tea or a stiff drink, have him sit in her finely decorated, two-story parlor, and offer a tour of the expansive grounds and her stable of cars behind the mansion. But this was different. She was not recruiting a fledgling young driver to her sports-car team. René may have been down on his luck, but he was a former top Grand Prix driver. He had driven for most of the greats: Bugatti, Maserati, and Alfa Romeo. He had been the teammate of many of the fastest drivers in the world. A fine house and a garage of dazzling shiny cars were not likely to impress him.

She knew she had to win him over. From the moment she entered the room, full of energy and enthusiasm and promise over what could be, René was swept away. Lucy did not so much ask him to join her team as tell him why he ought to. Jean François and Delahaye had proved they could build fast, reliable cars, and their design for the new formula was being given every attention. With her money, there was no resource that they or René would find wanting. "It is to be professional all the way," Lucy said. René would have a salary, his pick of other drivers, and the freedom to help develop and test-drive the new car.

Most important, together they could bring France back into the victor's circle and pierce the invincibility of the German Silver Arrows.

Neither recorded much about what was said that day in the autumn of 1936. Perhaps Lucy told René about always being the outsider looking to make good — not French enough in France; not American enough in America; the *nouveau riche* upstart in a class-conscious Europe; a woman in a sport ruled by men; a wife and mother who preferred the garage and the racetrack over the conventional hearth and home. No doubt she could tell him a thousand tales about malign looks and whispered comments.

Perhaps she called to his attention the fact that he was an outsider too. She knew he had been forced off the Ferrari team. She knew the Germans would never take him. All his skill and experience meant little when balanced against the name Dreyfus. It did not matter whether he saw himself as a Jew. They did — all the more so because he was the only one in the Grand Prix. Even if he did not hear the slights or see the looks, they happened nonetheless.

Together they could tip the scales in their favor, the outsiders atop the Grand Prix. Imagine it. The journey would not be easy. He would have to rediscover the fierceness of his early La Turbie days, as well as the fearlessness that first catapulted him into the Grand Prix with his 1930 Monaco win. Whether Lucy spoke of all or none of this, whether she needed to or not, René was bowled over by this fascinating lady "who talked a very good story." In the end, as usual, Lucy got her way.

Part III

8
Rally

AFTER THE SUMMER Olympics, the Mercedes team took a break from competition to work on the cars and get them ready for the Swiss Grand Prix on August 23. Throughout the 1936 season, they had struggled with their redesigned W25s. The engineers had increased its engine size while shortening the wheelbase. This gave the cars a striking look, but race after race they proved difficult to control and broke down frequently. "Sheep in wolves' clothing," one writer labeled them.

Bernd Rosemeyer continued his winning streak. At the Coppa Acerbo in seaside Pescara, Italy, he trounced all comers, reaching speeds of 180 mph on the straights. He also claimed the Freiburg hill climb, a 127-turn, tree-lined ascent that was one of Europe's most difficult events. His teammate Hans Stuck was usually the favorite to win, but vicious posters hung around the town that trolled him for being married to a Jew had thrown him off his game. "A Jewish slave. No start for Stuck," they read.

In the run-up to the Swiss Grand Prix, reporters were calling Rudi Caracciola a driver whose time has passed. The old lion must give way to the new, they wheedled. Rudi intended to prove them wrong, and he thought the adjustments to the W25's handling would give him an assist.

In the first two laps of the short, diamond-shaped Bremgarten circuit, he held a narrow lead over Bernd who stuck to his tail a mere second behind. When Bernd tried to pass in the third lap, Rudi blocked him. In the fourth, Rudi angled his W25 in front of him again. A race official waved the blue flag, a signal to Rudi that a faster car behind was trying to overtake him. He refused to give way, unwilling to let the upstart pass.

Expecting a crash, the crowd held its breath and watched the two Silver Arrows flash around the circuit, feet apart, at speeds averaging over 100 mph. Over the next several laps, Rudi maintained his slim lead. More officials presented blue flags. Grimly, Rudi continued to block Bernd, hoping he would make a mistake. But his rival drove flawlessly. Finally, in the ninth lap, Rudi caught sight of the lead Swiss race official standing at the circuit's edge and waving his own blue flag. He looked like he might well walk out onto the track on the next go-around if Rudi persisted. Reluctantly, he gave way.

In the next straight, an enraged Bernd roared past. He finished the lap with a new record, then did so again, and then again. Rudi fell back, unable to maintain the pace. Before the halfway point, a broken rear axle forced him from the race, and Auto Union went on to yet another victory. Only one of the four Mercedes finished the race.

That evening at the Bellevue Hotel, Rudi was traveling down in the filigreed elevator on his way to attend the ceremonial dinner. The elevator stopped at another floor, and in strode Bernd and Elly. The two drivers, now in tuxedos, had not spoken since the race. From the look on his face, Bernd was still seething.

"Well, young man," Rudi said. "You did well. May I add my congratulations?"

Elly responded first, accusing Rudi of putting her husband at great risk. Bernd cut her off. "So you think I did well, do you?" he said, his voice rising. "I would have done a lot better if you hadn't been in my way for so long!"

The elevator arrived in the lobby, and the two drivers stepped out, now trading insults, while everyone who was gathered there looked on. Finally Rudi and Bernd separated. Throughout the evening, they exchanged glares, but stayed clear of one another.

On September 13, Bernd blew away the field at the Italian Grand Prix. With five major Grand Prix victories in only his sophomore year of competition, he had clearly dethroned Rudi as European Champion.

Reporters deemed him "a remarkable driver: dazzling, a true virtuoso of the steering wheel."

The season over, Rudi returned to his Swiss chalet to await the next season. Mercedes had hired a new engineer, Rudolf Uhlenhaut, and they charged him with replacing the W25 with a new car for the extended year of the 750-kilograph formula. When it was ready, Rudi would travel to Monza, the Mercedes winter training ground.

He wanted Baby to join him there. Alone. Louis would not be returning to the team in 1937. He had suffered an abysmal season with Mercedes; after overturning his car at the German Grand Prix, he had returned to Paris with Baby to convalesce from a shoulder injury. Her absence made Rudi realize how desperate he was to be with her. Although Louis was one of his closest friends, Rudi believed that having Baby as his wife would be worth the betrayal.

That fall, after he had recovered from the season's strain on his leg, Rudi traveled to Paris to propose to her. They clearly cared for each other, he said. Louis was too much the playboy, still resisting marriage, he said. She deserved someone who would be committed to her forever. But Baby said no.

Into the winter, a lonely Rudi read often in the newspapers and magazines about Bernd Rosemeyer and his new bride—the darlings of the social columns. He read how they needed police escorts to part the crowds wherever they went. Then he read about how they had flown together to a race in Cape Town, South Africa, and to Victoria Falls in Southern Rhodesia afterwards. Bernd was even training for his pilot license. Reporters spotted them partying with Max Schmeling in Berlin, then skiing in the Alps. The celebrity couple was everywhere. In the sports pages, which were looking ahead to the 1937 season, Rosemeyer was the pundits' favorite. Rudi decided that he could not stand for that, nor could he live without Baby for another year.

"What are my plans for next year?" René repeated the question from *L'Intransigent*'s Georges Fraichard. "I will run next year, you can be sure."

"In which car?" Fraichard asked. It was October, the "season of rumors," and the two men were conducting their interview over breakfast in Paris.

"It's still a secret . . . But what I can say is that the contract binding me to Talbot has expired."

The interview ended without René revealing any more. Although he had committed to driving for Lucy Schell, he questioned the wisdom of his choice. Yes, Delahaye was performing very well with its sports car. Charles Weiffenbach and his engineer Jean François clearly knew what they were doing. The company was weathering the strikes in France better than Bugatti or Talbot. At the Paris Salon, their touring models, which were based on the 135, were lauded as the "most beautiful of the beauties" and were selling fast.

Yes, too, Lucy was bringing her fortune to the design of a formula car. Undoubtedly, she aimed to win. René liked her brashness and unswerving passion, and, as he later wrote, her idea to run a team in France with Delahaye, like Ferrari had done in Italy with Alfa Romeo, was, "I thought, ambitious. But what she said made sense."

Nonetheless, he worried that he had been sold a dream, a very impractical one. Delahaye was essentially building a Grand Prix car from scratch. Any such venture would have teething trouble — or worse, need significant overhaul. Which would take time to resolve. Then there was the question of their race organization and pit crews. Neubauer had proven how essential these were for Mercedes. Lucy had no experience on the Grand Prix circuit.

As the weeks passed, doubts gnawed at René. He feared he had abandoned Talbot prematurely. Tony Lago might well resolve the issues with the T150. Or perhaps he could have tried Bugatti again. Jean-Pierre Wimille had enjoyed some success with them that past year, enough to become the best-ranked French driver, even though this accolade did not even place him among the top ten in Europe. René knew that he was in no position to boast. He was absent altogether from the French list.

Despite his misgivings, there was no turning back, nor any other choice. To return to the Grand Prix, he would have to take the leap with Lucy Schell. If the venture was to lead to success, he knew that he needed to prove to himself that he was still good enough. He resolved to try.

On December 10, 1936, a few days after the definitive AIACR ruling on the new formula was finally made, everybody in the motorsport world heard about René's decision when Charles Weiffenbach and Lucy Schell announced to the press the formation of Écurie Bleue.

The team would represent Delahaye in all sports-car competitions the following year and would field a "horse" in the 1938 Grand Prix, Weiffenbach declared. Lucy Schell, he continued, would be the "boss" of the whole affair and René Dreyfus its "premier driver." He concluded: "The story of racing is the story of commerce, of industry, of life itself."

As for the design of their car, Monsieur Charles was coy. There was so much to do before the new formula season, but they thought they had over a year to get it right. Then, on New Year's Eve, the French government made a surprise announcement that drastically accelerated his schedule to ready Lucy's race car.

The Million Franc prize. One million francs. It was a round number, elegant and long, one that sparked immediate interest and played well across newspaper headlines. On December 31, the minister of public works, Albert Bedouce, an old political hand, publicly declared that the fund, raised by adding a ten-franc fee to all driver's licenses and buoyed by sponsorships from automobile-related industries, totaled 1.48 million francs. The goal, Bedouce stated, was to inspire and support French efforts to build a car for the new Grand Prix formula and to reclaim their standing in motor racing.

The fund commission now devised a contest, open to any French auto manufacturer. In fact, after much deliberation and political wrangling, the commission had created two separate contests. The winner of the first would be awarded 400,000 francs for completing the fastest sixteen laps—200 kilometers—from a standing start at the Montlhéry road circuit by March 31, 1937. The winning car needed to meet the new formula standards and average a minimum of 146.5 kph (91 mph). This figure represented a 2 percent increase on the record average that Louis Chiron had established to win the 1934 French Grand Prix.

The percentage increase looked like a muddled bureaucratic compromise, especially given that the Silver Arrows and the latest Alfa Romeos would have had an easy time surpassing that average speed.

Further, competitors for the prize could surpass the engine limit

(3.0-liter supercharged engine or 4.5-liter unsupercharged) by 10 percent. Many scratched their heads at that allowance. The organizers explained that it had been instituted because of the limited time before the March deadline for manufacturers to meet the formula standard. However, given that Bugatti's T59 ran a 3.3-liter supercharged engine (an exact percentage match), it was evident that the initial prize was nothing other than a gift for the Molsheim firm. No other French manufacturer would be able to field a car in time.

The victor in the second contest would claim the big prize: the 1 million francs. The rules matched the first, without the engine allowance. Deadline: August 31, 1937.

Days after the announcement, Talbot and SEFAC threw their hats into the ring to win the "Million Franc Race." Lago told reporters, "I will try my luck." Ettore Bugatti diplomatically stated that he would have to "adapt in a short time" to attempt the first contest. As for the second, he was more direct: "One thing is certain," he said. "I will go to battle. I will prepare for it, and I will endeavor to win it."

As for Delahaye, Monsieur Charles was against the minimum average speed, preferring the contest to be about the best performance. He thought the August deadline was restrictive.

"You are decided, nevertheless, to compete for the Million?" *L'Auto* asked.

"Perfectly," Weiffenbach said. "Delahaye will surely be on the line. We are already working on the problem."

The Million Franc Race had begun. As one historian enthusiastically wrote, "It was to catch the imagination of the French public as no other motor racing event had succeeded in doing, either before or since: a drama on a national scale playing on some of the deepest human instincts and emotions — passion for sport, admiration of skill and prowess, love of country — and all brought to the boil in a cauldron fired by that most magical of all motivators, lust for money."

Shortly afterwards, in mid-January 1937, René headed to the German port city of Hamburg. Lucy wanted him to make his debut for Écurie Bleue in a competition he had never run before: the Monte Carlo Rally. He would partner with Laury in one of their Delahaye 135s.

Initially René was reluctant. He was a race car driver, not a rallyer.

They were distinctly different breeds. One did not put a miler in a marathon. But after consideration, he thought that the whole affair would be tremendously fun. The average speed of 25 mph over the route seemed like it would make for an easy "long walk" over the 2,300 miles to Monte Carlo, especially since he was accustomed to maneuvering through packs of cars at quadruple that speed on narrower courses. He would also see some incredible scenery and enjoy the camaraderie of competing with Laury, whose love for cars exceeded even his own.

In Hamburg, René took the wheel of the canvas-topped Delahaye 135, and he and Laury headed to their starting point in Stavanger, Norway, a journey of 1,000 miles. René hoped to get a good feel for the car on the way. They had barely reached the Danish border before they were turned back by a wall of snow. A blizzard raged across northern Europe, making roads impassable. After retreating back to Hamburg, they boarded the cargo ship *Venus* to reach the west coast of Norway. Halfway there, René was curled in his seat, face blanched, stomach roiling like the deck under his feet. Outside, a dry, icy wind howled, and foam-flecked waves slammed against the hull. All of the North Sea looked like it had risen up to send him to its black depths. Ships were

René Dreyfus and Laury Schell at the 1937 Monte Carlo Rally

sinking or foundering up and down the Norwegian coastline, and at one point theirs needed to stop to assist a vessel and take on its passengers.

Laury Schell, who had the constitution of a grizzled sailor, thought that René needed to buck up. He promised him that there would be more trouble ahead once they reached firm ground. By the time they arrived in Stavanger, they rushed to make the start. Neither had slept or eaten in almost a day. Still, at 1:00 p.m. on January 26, sporting a fur-lined black leather trench coat, René was excited to be at the starting line. Everybody—from the organizers to the local people to the thirty driver teams—was friendly and enthusiastic, particularly since the sun had come out.

René made quick work over the first hour, averaging 40 mph. He liked how the little Delahaye drove. What its engine lacked in power, it made up for in quickness; the brakes were tight; and its steering was responsive. The car just felt right.

Then the road slickened into a skating rink, one buffeted by ferocious winds. René slowed, but he found that the 135 persisted in shearing out from underneath him, something he had never experienced while learning to drive around the sun-kissed streets of the Riviera.

He and Laury put chains on the tires, but the road ahead only got worse. On the first leg to the control point in Kristiansand, a 185-mile route through a rugged, pine-strewn landscape skirting the Norwegian coast, they passed several competitors stuck in snowbanks. Most were behind schedule. Two decided to abandon the rally altogether.

The sky darkened early as René was attacking a high pass. White-knuckling the steering wheel, he struggled to keep the Delahaye on track. The sharp slalom of the descent proved even trickier, but he avoided calamity. They reached the Kristiansand checkpoint a few minutes ahead of schedule. Eleven of their fellow rallyers lost points for their late arrival.

After a brief break, they returned to a road that looked like it had been carved into a corridor of snow. Its hard-packed walls rose above their roofline. Any loss of control, and it would be akin to crashing inside a concrete tunnel. René had gained a newfound respect for Lucy's sporting achievements.

Sixty miles out of Kristiansand, the 135 threw a chain, which then twisted around the brake drum. René pulled off the road, close to the corridor wall, to fix the problem. As he and Laury labored in the middle of the night to loosen the chain, the Arctic cold numbed their hands and bit their faces. Ten minutes passed. Then half an hour. As they struggled to untangle the chain, competitors slowed to curl around them on the narrow road. Time was running out. Finally, they cut it free with pliers.

They had fallen almost an hour behind, time they tried to recover over the next 150 miles to Oslo by barreling onward, tires aimed at the grooves carved into the compacted snow by the other competitors. Whenever their wheels abandoned these "tramlines," as they called them, their tail end performed acrobatics.

They arrived six minutes past the control-point deadline and incurred three penalty points. Barring misfortune hitting every team in the race, they had forfeited their chance at a first-place finish. Such were the vagaries of the Rally.

Laury and René continued on into Sweden, then down the coastline toward Denmark. René learned to trust the Delahaye and to accept the risk of an accident that came at every turn and downhill sweep of the road. If they careered into a snowbank, they simply shoveled their way out. With each hour, he grew more confident.

René was much less assured when it came to navigation. Ice and snow blanketed most signs, and he took several wrong turns while Laury was sleeping that were discovered only after he woke up.

At one crossroads, René had lost all confidence in his sense of direction. He elbowed Laury to ask for help.

"Which way do you want to turn, René?" Laury asked.

"Left."

"Turn right, René."

This scene repeated itself often. Whatever direction René chose, Laury opted for the opposite, and they began arriving at their checkpoints easily on schedule.

Forty hours after leaving Stavanger, they got a couple hours of uninterrupted sleep on a Danish ferry. Then it was back to the road. Bedraggled, yawning every few minutes, René continued to drive, wisely coached by a co-driver who was running his eleventh Monte Carlo Rally. "Bet that today you would prefer to be on the starting line of a Grand Prix race," Laury joked as they drove down a slick stretch of French highway where several trucks had slid off the road into ditches.

René agreed. In some ways, the rally was far more dangerous, particularly since that year they were competing during one of the worst winter storms in decades.

Belying their travails, they arrived in Monte Carlo on a hot, clear day that might even have been mistaken for summer. René and Laury performed well in the series of acceleration and braking tests. They finished fifth overall. Another Delahaye driver, René Le Bègue, won first place.

Some called the sixteenth running of the Monte Carlo Rally the hardest ever. Only half the drivers who left Stavanger made it to Monaco. Lucy was well versed in the trials her new driver had faced, and after the rally, she invited René to her expansive villa beside the Jardin Exotique de Monaco, high in the hills above the sea. There, with Laury, they shared stories of the *chaine maudite* (the damned chain) and other hair-raising moments. Relaxing by the Mediterranean, René savored those adventures, but it had dawned on him just how hard Lucy intended to push him in the Grand Prix.

In February 1937, before Rudi went to Monza to trial the W125, the latest Mercedes formula car, he traveled to Paris to see Baby. A recovered Louis was away skiing in Austria, but all reports were that he would not be joining any team for the season.

Again Rudi proposed. They were well suited, he said. He promised the kind of life she wanted, one of stability but with the thrill of motorsport in the mix too. He would build her a grand villa on Lake Lugano in Switzerland. They would pass the rest of their lives together.

Again, Baby refused him. She believed that Louis would marry her if he knew that losing her was his only other option. Bereft, Rudi left for Italy.

At Monza, Mercedes engineer Rudolf Uhlenhaut presented him with a marvel. In only six months, Uhlenhaut had reinvented the Mercedes formula car, a surprising accomplishment for anyone and particularly so for a thirty-year-old mechanical engineer of reserved demeanor, little management experience, and a preference for staid tweed jackets.

When he was hired as technical director for the Mercedes racing department, Uhlenhaut inherited a team of 300 engineers, technicians, and mechanics, most of whom were dispirited after the performance of their

1936 design. That autumn, to plumb the problem, Uhlenhaut had taken a pair of the supercharged W25s out for a spin around the Nürburgring. To keep the excursion quiet, only a few mechanics joined him.

Uhlenhaut had never driven a race car, and he started out at a snail's pace, but after a thousand miles of driving over the next few days, often at top speeds, he had a better understanding of what he needed to do. To begin with, he replaced the box-section chassis, which warped and vibrated on uneven roads, with a stronger, oval tubular frame made of a nickel, chrome, and molybdenum steel alloy. He lengthened the wheelbase by a foot to provide stability. He improved the brakes and overhauled the front and rear suspension, softening the springing and providing much better traction at high speeds.

As for the engine, Uhlenhaut stayed with the straight-eight but increased its capacity significantly, to 5.66 liters, and improved its crankshaft. In tests at the Untertürkheim plant, the engine developed an incredible 589 horsepower. To stay under the 750-kilogram maximum weight, most of the car was built with advanced light-alloy steels.

For several weeks, Rudi trialed the W125. The wheels clung to the road, and it was fast—faster than anything he had ever driven. He reached 88 mph in first gear. In second, 137. In third, 159. In fourth, the top gear, he accelerated to 199 mph. He was elated by these results. At last he had a new weapon against Bernd Rosemeyer and Auto Union.

While Rudi was trialing the new Mercedes in Italy, René and the Écurie Bleue team traveled south from Paris in two trucks, each towing a covered trailer with a Delahaye 135 on board. The sides of the trucks were painted with the team mascot, a British bulldog. Lucy owned a pair of the breed and was mad about them.

The team was headed to the Pau Grand Prix. Again, no German or Italian teams had entered the race: the ACF had decided for yet another year that only sports cars could run in events in France, and in protest these foreign teams refused to participate.

Jean-Pierre Wimille was the obvious favorite, especially after he posted record practice times. His Bugatti T59 had been built for the Grand Prix, but the supercharger was stripped out to meet the sports-car regulations. René's best practice lap was a dismal five seconds slower.

On race day, February 21, a wet afternoon, Jean-Pierre took the lead and never lost it. René finished in third place, a minute back.

Next, René competed in the Mille Miglia, another first for him. Rudi Caracciola maintained the distinction of being the only non-Italian driver to have won the 1,000-mile Italian race. During the first half of the race, conducted in a pounding April rain, René was running well. He was in second place and only thirteen minutes behind Ferrari driver Carlo Pintacuda. Over such a long race, that was a surmountable gap. René was closing on the Italian, on a stretch of the course near the Adriatic Sea, when he came to a mountain bend. René cut too close to the edge of the road, and his tires spattered hunks of mud across his windshield and visor. Temporarily blinded, he shot off the road, hit a rock, and flipped his car. Thrown clear, he was fine apart from some bruises and wounded pride. Some farmers hauled his car back onto the road with a pair of oxen, but he had to retire from the race.

René's close run on the leader and a third-place finish by teammate Laury Schell were the best showing by a French team in the Mille Miglia's history. Overaggression in the mountains probably foiled the race for René, but if Lucy was any example, better that than timidity.

On March 27, lingering in the glory of his Pau Grand Prix win, Jean-Pierre Wimille drove the Bugatti T59 again, this time with its supercharger returned to its rightful place, in an attempt at claiming the first tranche of the Fonds de Course money. He had a good start at pacing the Montlhéry road circuit at better than 146.5 kph average, but mechanical trouble forced him to stop before the required sixteen laps. He suffered the same in a follow-up run the next day. Then, proving just what a fix the competition was, the Fonds commission extended the March 31 deadline — in recognition of Ettore Bugatti's "past great efforts."

When reporters asked Weiffenbach his opinion on the extension, he demurred. Tony Lago was less diplomatic, stating that the commission should just give Bugatti the 400,000 francs and be done with the charade of equality. Lucy did not express her thoughts to the press, but the commission's efforts on behalf of Bugatti were yet another example of the entrenched old boys' establishment that defined her sport — and world — and it left her numb.

On April 12, the Bugatti team tried again. Jean-Pierre made a good standing start and managed the first lap in five minutes and seventeen seconds, only a ten-second-slower pace than he needed to maintain throughout the sixteen laps. He made up this lost time and finished the 200 kilometers with an average speed of 146.7 kph in one hour, twenty-one minutes, and 49.5 seconds. It was a few blinks of the eye (4.9 seconds to be exact) under the baseline time limit.

"Bravo, Jean-Pierre," headlined *L'Auto*. Asked if he had felt any fear about how hard he had to push his Bugatti to beat the speed target, Jean-Pierre cockily answered, "In no way."

Over the course of 1936, he had bested René with his precise, emotionless driving style. He never missed the opportunity to needle René about his poor performance while he was driving for Talbot, and the two were becoming heated adversaries.

Now Jean-Pierre was popping champagne corks over his Fonds win —dubious though it was—and boasting that the million francs would be his by August.

René aimed to snatch the prize away before then. Monsieur Charles and Lucy were no less determined to beat Bugatti to it. But their Grand Prix car remained unready.

Throughout April, Jean François continued to struggle with building his new engine. Early in the new year, their foundry had cast the magnesium-alloy cylinder blocks, but during the process these had developed gas bubbles. When they cooled, their skin was porous and they failed to hold water under pressure. They looked like a hill of termites had burrowed into them. Delahaye was experienced in casting blocks in iron but new to working with magnesium.

In late spring, François succeeded in assembling an engine. Then, during runs on the testbed, other issues arose. Again the magnesium blocks presented problems, this time because the alloy expanded and contracted at a different rate to the steel in the studs that held it in place. The engine tore itself apart when heated up.

Dispirited with his progress, especially after Bugatti claimed the initial Fonds prize, François considered abandoning the whole design. When Monsieur Charles failed in his gruff efforts to spur him onward, Lucy offered her own: work until it works.

9

The Winged Beetle

I N MAY, RENÉ found himself back in a Maserati, this time at the Tripoli Grand Prix, positioned at the rear of the starting grid. The red 1.5-liter *voiturette* looked like a toy car compared to the formula Grand Prix models at the front piloted by the Germans and his former Alfa Romeo teammates.

On weekends when there were no sports-car races scheduled, Lucy permitted René to run in *voiturette* class events as an independent driver, and Maserati provided him with one of its cars for several competitions during the 1937 season.

In Tripoli, the organizers allowed *voiturettes* to compete to help fill out the total field of thirty, even though the small cars had no chance against their bigger cousins. René was one of eleven drivers in his class, most of whom were much younger. The *voiturette* class had its own victor, though the prize was one-third the amount awarded to the overall winner. Regardless, participating was a far cry better than sitting in Paris listening to the event on the radio.

It was over 100 degrees on the Mellaha track when Governor Balbo lowered the Italian flag. The formula cars, a mix of Silver Arrows and Alfa Romeos, were long gone before René, whose Maserati was sluggish at the start, moved into the first high-speed turn. Before he finished the lap, a thick haze of churned-up sand obliterated any sightlines.

The Maserati was a nimble, punchy little animal, and by the halfway point René led his class. Over and over, the formula cars screamed past him through the fog of sand, seemingly out of nowhere, sometimes at speeds of over 175 mph. Inhabiting the same track with these beasts was terrifying. Nevertheless, René came first in his class, showing the same skill and steeliness that had made him a top Grand Prix driver in the past.

The Mercedes W125s swept the field, the first time the new model had been raced. Hermann Lang, a twenty-eight-year-old former Mercedes mechanic, won for the team and was the star of the celebration at Balbo's palace, much to the chagrin of Rudi and Manfred von Brauchitsch. A year before, Lang had been servicing their cars. When the team grabbed a drink at the bar, Manfred ordered "champagne all around . . . Oh, and a beer for Lang." Nobody paid much attention to René, nor to his class win in the Maserati.

Between sports-car and *voiturette* races, René often visited the Delahaye factory to check on the 145. Jean François kept tinkering with the engine until every cylinder leak, loose valve, and structural weakness had been fixed. The magnesium block weighed half of what one cast in iron would have weighed. So, though it tipped the scales roughly the same as the 135 engine, the V12 provided 50 percent more power and turned 1,000 rpm faster.

In early June, René received a call from Baby Hoffmann. She wanted him to come over for dinner, alone. It was an odd invitation, particularly because she and his wife remained good friends since their Scuderia Ferrari days together. René arrived to find Baby stuffing her suitcases. "I'm leaving Louis," she said. "I'm going to marry Rudi."

René made no effort to convince her to stay. It was her life. Neither could he give her the approval she wanted. Louis was one of his closest friends. Further, although René did not believe Rudi was a Nazi sympathizer (unlike Manfred), he had clearly bent under pressure to be one of its foremost, Hitler-saluting heroes. Instead of pointing this out, René simply drove her to the train station.

Early the next morning, Louis hammered at his door. René answered, and his friend stormed into his apartment. "How could she do this to me?" Louis demanded.

René failed to calm him. Louis was beyond reason and forgetful of the fact that he had stolen Baby from her husband years before. He left in the same rage in which he had arrived.

Death, rivalry, betrayal, and politics—together they had irrevocably riven apart the Grand Prix world.

On June 19, Rudi and Baby wed in a quiet ceremony in Lugano, Switzerland. Photographers snapped the newlyweds on a balcony overlooking the lake, their smiles irrepressible. It was the happiest Rudi had been in years. Three days later, they boarded the German ocean liner *Bremen* to travel to the Vanderbilt Cup in New York.

In the opening months of the season, Rudi had found little success in the new W125. In Tripoli, he had been slowed by a faulty supercharger and some interference with his former teammate Luigi Fagioli, now with Auto Union. At AVUS on May 30, Mercedes and Auto Union fielded cars with souped-up engines and aerodynamic bodies for the non-formula event, which was attended by Goebbels and other Nazi high brass. In the long straights, their rockets reached over 240 mph. Rudi suffered engine failure, and Lang won again.

Then, on June 13, Rudi finished a distant second to Bernd Rosemeyer at the Eifel race. After the race, the young driver returned the swivel stick that Rudi had given him in 1935, with an admonishment. "Well done, my dear fellow," Bernd said. "But in the future don't just drive round and round the circuit; use your head."

Rudi was looking for revenge, and he hoped to find it at the Vanderbilt Cup. The race, founded by the wealthy railroad family at the turn of the century, and boosted by big cash prizes, was intended to develop motorsport in America. World War I brought its first run to an end; then, in 1936, another Vanderbilt scion restarted the event. Interest blossomed after Nuvolari's meteoric win in 1936.

For the 1937 race, Adolf Hühnlein sent both Mercedes and Auto Union to compete. Victory there would boost exports and, more important, provide a huge propaganda win for the Reich.

During their five-day voyage aboard the *Bremen,* which the Nazis promoted as the world's most advanced high-speed ocean liner, Rudi and Baby rested by the pool, shot clay pigeons from the deck, played billiards and deck games, watched films, gambled, and dined extravagantly. All the ship was abuzz with the news of their impromptu marriage. In a temporary truce, Bernd and Elly gave the newlyweds an antique pewter mug. They were rarely left alone, particularly since both German teams were traveling with the usual entourage, there to support their drivers: Rudi and Englishman Richard Seaman, driving for Mercedes; and Bernd and Ernst von Delius for Auto Union. Two high-level Nazi officials were also present — Jakob Werlin and Dr. Bodo Lafferentz — as well as several SS "minders."

One night the ship's captain invited the racing parties to his table. Nearby sat a large Jewish family. Lafferentz, who led Strength Through Joy, a Nazi state-operated leisure organization, blustered loudly, "In future, Jews must not be allowed on any German ship."

The captain stood up to him. "That is a family of ten, from Hungary. They travel on my ship at least twice a year, and they are most welcome."

The table went quiet, as the thoughts of those present turned to what the SS would do to the captain once he returned home.

Days later, the *Bremen* moved into New York Harbor, past the Statue of Liberty, under a pouring rain. As the team disembarked and the Silver Arrows were unloaded, a few protesters on the docks shouted "Nazis!" and threw rotten cabbages at them.

The ugly scene failed to mar the adoring reception they received in the city, especially from the press. Such were the crowds that team officials hired bodyguards from the Pinkerton Detective Agency. The spotlight focused mostly on the young Auto Union champion and his "aviatrix" wife. "Hello Bernd!" read one headline. Noted American columnist Bill Corum wrote, "This Rosemeyer family is in a deuce of a hurry. He drives at almost 250 mph on an autobahn, she flies almost as quickly around the world; what will happen one day if they have a son? There will be nothing left for him but to visit Mars in a rocket." Little did Corum know, but Elly was pregnant with their first child.

At the Roosevelt Raceway in Long Island, reporters and fans marveled at the Silver Arrows. They had never seen cars flash by at such speeds, nor make so much thunderous noise, nor tear around the narrow, windy course like they were running on rails.

American champion Rex Mays, in an independently run Alfa Romeo, was out of his depth. Early in the race, Bernd and Rudi jockeyed for first place. On the tenth lap, Bernd blazed down a straight at 158 mph to catch up with Rudi and barely slowed going into a sharp bend. His P-Wagen looked like it might fling itself off the track, but somehow he drifted perfectly out of the turn into the lead.

In the end, Bernd easily won the race and the $20,000 purse. At the prize-giving, the grand doyenne Margaret Emerson Vanderbilt mooned over him. Face blackened with grime but for the outline of his goggles, he looked a true conquering hero. After she handed Bernd the oversized silver trophy, she said, "Sir, I think you're just grand and your motor car is *wunderbar*."

The *New York Times* remarked that the German cars made the race the "most spectacular automobile marathon ever witnessed in this country." Only the disappearance of Amelia Earhart over the Pacific garnered more coverage.

The win was a publicity bonanza for the Third Reich. Hühnlein ensured that the German press back home parroted his press releases about the triumph in America. Lines to be quoted included: the "swastika flag hung on the victory mast at the raceway for over four hours"; the American drivers were "motor-cowboys" while the Germans were considered "gentlemen" who showed "impeccable order and discipline that left a lasting impression"; and the Silver Arrows showed undeniably that "German technology" was superior to anything in the "New World."

There was even a brief report that some New Yorkers had scrawled well-wishes on the boxes of spare parts the teams brought to the track. "We hope that you win! Heil Hitler!" read one. "Show the Americans what German racing cars are! Heil Hitler!" read another.

When the Mercedes and Auto Union drivers returned to Berlin, huge crowds greeted them. Bernd was promoted to SS captain and feted by Himmler. Rudi avoided the attention.

On the afternoon of June 25, while the German teams were crossing the Atlantic, the Delahaye 145 arrived at Montlhéry for the first time. Lucy Schell called no press for the event. There was a chance that her self-financed new Delahaye might sputter and die — the gearbox might break, or the tires might even fling off — if recent experience in the 135 was any guide.

Over the past couple of months, her team had been bedeviled by accidents and mechanical failures, leaving Bugatti and Talbot to dominate the French sports-car races. One of the few bright spots was the tenacity of her team captain.

At Le Mans, the week before, René had fallen back to the middle in the twenty-four-hour endurance challenge because his co-driver had bizarrely broken the door during a pit stop and the fix had taken almost an hour. Telling Monsieur Charles that he would claw his way back into position by racing "as if this were a Grand Prix," René proceeded to pilot his 135 at a delirious pace for ten hours straight, much of it in the dark, through the wheat fields lining the course. He had placed a remarkable third.

The Écurie Bleue transporter rumbled into the Montlhéry autodrome. Lucy was there to greet it, along with René and Jean François. Weiffenbach and some board members showed soon after. Besides this small crew and a couple of track officials, the grounds were empty.

The 145 rolled down the transporter's ramp onto the concrete track. There were no cheers, no champagne corks, barely even a murmur. Those who had never seen the car at the factory were shocked by its appearance. The 145 looked nothing like its trim, elegantly shaped predecessor, nor like any other Delahaye they had seen before. Outfitted with mudguards mounted high above the wheels for the upcoming sports-car-only French Grand Prix, it looked even more an anomaly.

"Gone was the classic shield-shaped radiator surround," a Delahaye aficionado later observed. "In its place was a broad blunt snout . . . Gone, too, was the slim tapering profile of the old 'Six'; the V12 body maintained the same ample girth all the way from the front to the back of the cockpit, its sides thereafter drawing in only slightly toward the flattened tail. The car seemed to hug the ground, its extraordinary low build making it look even longer and wider than it was."

Some thought the 145 must have been inspired by Lucy's bulldogs. Or an electric lightbulb tilted on its side. Or the "pug-nosed dumdum bullet." One critic figured that the inclusion of the mudguards in the design would appeal only to an etymologist keen on praying mantises. It was called "weird," "brutal," and "downright ugly." René thought it "the most awful-looking car [he] ever saw." Aspersions aside, its shape was utilitarian, from the bulbous nose that housed the extra engine width to the elevated mudguards that reduced drag.

All that mattered to Lucy was that it ran — and fast.

Presentation of the Delahaye 145, mudguards and all,
in the summer of 1937

Jean François insisted that he first test the 145 on the track. Lucy might have written the checks—and alternately cajoled and threatened him to push forward—but the car was his creation after all. As he tucked himself into the cockpit, incongruously dressed in a sport jacket and tie, the intrepid Georges Fraichard arrived with a photographer. He must have been tipped off.

"How is she going to behave?" the *L'Intransigent* reporter asked.

"She has never been driven before," François replied.

His mechanics started the engine. There was no deafening yowl as the V12 took life. It was more of a sharp, pulsating rasp that broke the monastic quiet on the track. As François drove away, switching through the gears, the pitch of the engine grew deeper, more authoritative, like it was settling into its own distinctively aggressive voice.

René smiled as François thundered past. Any thoughts about the car's ugliness vanished. Turning to Fraichard, René whispered, "She's pretty, is she not? She rides low and gives a beautiful impression of power."

When François finished a couple of laps of the autodrome, he gave over the pilot seat to René. Envious of his former Grand Prix colleagues competing in the Vanderbilt Cup, he knew that the 145 was his only path back into that arena. Before he set off, he pulled his white linen cap down onto his head and secured his goggles. On the first lap, he went slowly, the steering wheel vibrating from the power of the engine. Then, as he began to get a feel for the car—the stiffness of the reinforced chassis, the almost featherlight gear change—he pressed his foot down all the way on the accelerator.

The Delahaye leaped ahead and banked around the autodrome, steady and sure as a ball just launched around a roulette wheel. René covered a lap in 126 mph, then did several more, never exceeding 4,000 rpm. Finally he pulled up beside the huddle of onlookers and stopped sharply. The brakes were powerful too. There were congratulations and handshakes all around. Lucy was very pleased. Her car ran. It ran very well indeed.

After the celebrations, René returned to the track to put the Delahaye through its paces. A half hour later, the engine overheated. François returned the car to rue du Banquier and made some refinements, notably expanding the air scoop at the front. Then the car went back to Montlhéry, and René tested it again. More trouble with ventilation. It was back to the factory to cut louvers in the hood. More tests. More problems. More modifications. They were on a rushed schedule—perhaps an impossible one.

On July 4, when the 145 returned to Montlhéry for the French Grand Prix, many questioned whether it was a race car masquerading as a sports car—or, once stripped of its road gear, vice versa. Monsieur Charles and François replied that it had been built for both events. Versatility was a hallmark of the overall design.

A slate-gray sky looked down on a dismally attended event. The French Grand Prix, in its second year as a straight sports-car event, was a shadow of its former self: a meager eleven entries, no foreign-built cars, few spectators, and disdain from the international motorsport scene, particularly Adolf Hühnlein, who chided, "Some countries would like the engine of the race car to be abandoned in favor of that of the sports car. We will not let ourselves be dragged down this path."

Scoffs and bemused gazes welcomed René when he brought the Delahaye 145 into the pits. *Autocar* reported that the new car looked like a "winged beetle" with its headlights and elevated mudguards. Few of them knew what to expect of its performance. Even René had his doubts. Since its first test run two weeks before, he had worked almost every day with François to ready it for the race. Unresolved problems persisted.

"Obviously, it would be miraculous if it did well," Weiffenbach told a huddle of journalists before the start. "But for all those who stayed up all night to finish it before the Grand Prix, they deserve that we at least participate."

By the second lap, René already knew there was trouble. The engine was off-pitch. One of the cylinders was misfiring. He was losing speed. Moments later, he came into the pits and climbed out, shoulders sagging. The mechanics rushed to replace the twenty-four spark plugs. René returned to the circuit, far off the lead. Next, flecks of oil spat from the engine. Something significant was wrong. He returned to the pits after the seventh lap. More new spark plugs. The 145 limped back onto the track. The oil pressure plummeted. He barely made it back to the pits to abandon the race.

Louis Chiron won in a Talbot. He had joined Lago's team shortly before the race, and it was his first competition since his accident on the Nürburgring in July 1936. Talbot took second and third place as well. René wondered again if he had left the team too soon, replacing its early stumbles with the same ones he now endured with Écurie Bleue. But he knew there was no turning back. He had chosen his horse and would now have to ride it. Notably, in his best lap at Montlhéry, René finished thirty seconds behind the pace he would need on the same circuit to have a chance at "the Million."

Two weeks later, René skirmished with Jean-Pierre Wimille in the first laps of the Grand Prix de la Marne. The Bugatti driver had won at Le Mans, cementing his reputation and further stoking his ego. In a postrace interview, he declared that he intended to win every race from that point forward. He certainly followed that strategy on the triangular circuit at Reims, setting off at a rapid clip on a track that was blistering from the hot sun. René hung tight, his 145 running well. In the third lap, only a couple of car lengths back from the Bugatti, he surged down a straight at 120 mph to overtake Jean-Pierre. Suddenly, the steering wheel jarred in his hands, and the back tail of his car swung to the right. A blown tire.

The Delahaye spun three times until the rear wheels struck the curb-stone. René was almost flung out of the car, the brace of his knees against the cockpit all that kept him inside. The 145 leaped into a neighboring field and then pirouetted at high speed several more times before grinding up a ridge to a halt. Dazed, surrounded in a swirl of dust, René crawled out. His arm felt wet. He looked down at his wrist. Blood was pouring from a cut, and he bound it quickly with his scarf.

When the dust settled, he inspected the blown front tire. Its tread was frayed, and the inner tube peeked out. His team had recently switched sponsorship from Dunlop to Goodrich, and their tires clearly weren't able to handle the heat of the track coupled with the weight of the 145.

René walked back to the pits, cradling his wrist. Unlike in accidents past, he was not unnerved. Rather he was mad at losing the chance to win. He went straight to the Goodrich representative and launched into him, saying that he had almost been killed because of his damned useless tires. How could they dare sell such junk? Before he could throw a punch, his brother Maurice interceded. "Shut up, René," he said. "You're not being fair."

Jean-Pierre won the day, and René and Écurie Bleue departed Reims for Paris.

Lucy detested losing, but she knew any competition that summer was merely a testing ground. She aimed to stake her ownership of the best French race car by winning the Million against the arrogant Bugatti boys and then knocking the Germans off their perch in the next year's Grand Prix formula season. Nothing less. Nothing more. The accolades the Nazi drivers and their cars won in America—and then exploited back home—fueled her intensity.

René doubted that they would accomplish the first of her ambitions. Delahaye's chances of achieving the 146.5 kph average before the deadline, or besting any increase over it made by Bugatti, looked slim. Against the Silver Arrows and their two leading drivers, Caracciola and Rosemeyer, who would boast speeds far surpassing anything the Delahaye could manage, René believed they stood no chance whatsoever.

On July 25, almost half a million fans lined the Nürburgring for the German Grand Prix. Tens of thousands crammed the grandstands and the Karussell, the banked switchback turn halfway through the course. One writer noted that "the whole of sporting Germany seemed to have descended" into the Eifel Mountains to watch the race.

The "cream of the European drivers" were there, competing in the fastest of cars. There was no mistake: the German Grand Prix had usurped the French as the season's premier event.

"Make a perfect start—a *perfect* start," Rudi repeated to himself from his position in the second row of the grid. Left foot on the clutch, he watched for the signal to go. On green, he stabbed the pedals and lurched forward, bursting through the first row, past Bernd, before they headed into the first curve, then Rudi dove down into the forest.

During the practice sessions, the Auto Union driver had astonished everybody by landing his own airplane on the track, climbing into a car, and then clocking the best times. Rudi intended to impress on the only day that mattered. After his quick jump off the grid, he ran a steady, cold-blooded race over the twenty-two laps.

Bernd was slowed by tire trouble and tried recklessly to recover his lost time. Ernst von Delius and Dick Seaman sparred down a straight at 155 mph and crashed terribly. Others bowed out because of mechanical problems, and Tazio Nuvolari simply could not keep up in his Alfa Romeo.

Rudi, sticking to his plan of two pit stops and an uncompromising pace, won with a long lead over his teammate Manfred, followed by Bernd. On the victory stand, Rudi raised his arm in a Sieg Heil salute when his name was called, then received the Hitler Prize with a smile. The trophy, a huge bronze bust of "The Goddess of Speed" with wind-swept hair and lightning bolts affixed to her temples, was quite a weight in his hands.

Bernd stood beside him on the podium, drawing heavily on a cigarette. Seaman would recover from his injuries, but Bernd was very shaken that Delius, his teammate and best friend, had been rushed to the hospital—and would soon die. Rudi assumed that Bernd's obvious despair was over losing to him again.

The next morning Hans Bauer, Hitler's personal pilot, flew Rudi and Manfred to Bayreuth, the Bavarian town where the Nazi leader had his country home. Goebbels welcomed them, and then there was a round of hearty handshakes with Hitler before a photographer took their pictures.

Later that afternoon, Bauer told Rudi that some party higher-ups had heard a rumor that he had taken Swiss citizenship; after all, he was building a house for his new wife there. Was it true? Rudi dug into his pocket and presented Bauer with his German passport. He was a patriot, Rudi implied, same as anyone at Bayreuth. He lived on Lake Lugano because the dry, warm air was good for his leg. This satisfied the pilot. Emboldened by his recent win, Rudi cheekily countered, "Now, I must ask you something. Do you suppose you could fly us to Stuttgart?"

After they landed, Rudi was paraded through the streets on the bed of a Mercedes truck that had been bedecked with flowers. At the gates of Untertürkheim, a banner read HEIL DEM SIEGER (HAIL TO THE VICTOR). A brass band and rousing chanting of his name welcomed him. At the party afterward, Wilhelm Kissel handed him a diamond-and-sapphire medallion fixed with a Mercedes star.

Rudi followed his German victory with wins at the Spanish and Italian Grands Prix, where he continued to drive the superb W125 with a mix of aggression, focus, and seasoned experience. At the end of the season, Rudi was named European Champion, toppling Bernd from his throne.

For four years running, the Silver Arrows had absolutely monopolized the Grand Prix, fulfilling one of the promises Hitler made at the 1933 Motor Show. Production figures at Mercedes and other manufacturers were growing by double digits every year, exports were rising, and profits were fat. The national autobahn project was also making huge strides. Hundreds of thousands of workers, fleets of trucks and machinery, tons of iron and steel, and enough concrete to fill 100,000 railroad cars—all went into creating 4,287 miles of "Hitler's Highways." The Führer was also moving forward with his dream of putting every German family in their own automobile, with the Volkswagen (the People's Car), a project spearheaded by Ferdinand Porsche. NSKK membership was on the rise as well.

Victory in the Grand Prix was the pinnacle and the inspiration for all these efforts to motorize Germany, and Rudi was once again at its apotheosis, the hero of the new Germany. He was constantly lauded as

"Caracciola, the man without nerves," and his every victory was played out in newsreels, his arm always raised in salute to the Führer. Never was he seen limping on his shortened leg.

He published a best-selling memoir, *Rennen — Sieg — Rekorde!,* which set him up as an unlikely champion and praised the role of Hitler in reviving his country's fortunes in motorsport. "The driver fights for victory and honor," Rudi wrote, "and the law of fighters is to burn oneself out to the last spark." Further stoking the Nazi rhetoric, the Goebbels machine proclaimed him to be a "frontline" soldier in the "racetrack battle" aided by the "brave, small army of mechanics."

Rudi supported the party line by taking part in advertising campaigns, speaking at the annual motor show, and appearing beside Hitler and other high officials at public events and private parties. He was, de facto, a standard bearer of the Third Reich, a regime whose growing military footing gave teeth to its veiled threats of "total war."

10

"Le Drame du Million"

I N LATE JULY, Lucy and her Écurie Bleue team established base camp at the Montlhéry track. They brought a mechanics' shop of equipment and spare parts and a stack of Dunlop tires—everything apart from sheets and beds. They were staying until they won "le Million."

Lucy wanted René to practice at Montlhéry until he knew the 12.5-kilometer course, including the road section and the autodrome, like his own bedroom in the dark. While Jean François was at the factory, building a new iteration of his V12 that capitalized on what they had learned from its performance throughout the season, René was at the autodrome with a 145 stripped to its race car essentials and running an earlier version of the engine. Its body was a thin, unpainted aluminum shell that looked like it had been hammered out by a half-blind panel beater. To reduce weight, François stripped out its second seat.

As August progressed, reporters flocked to Montlhéry to see who was practicing and who might make the first attempt. Would it be the fabled house of Bugatti? At the beginning of the decade, they had owned the Grand Prix, and they had already claimed the fund's first prize. Was this to be their comeback, championed by that dashing driver Wimille? Would Émile Petit, the noted engineer of SEFAC, produce a winner after a string of unfulfilled promises? Might Tony Lago, who talked a good game, build off his success at the French Grand Prix and field a Talbot contender, with Louis Chiron, three-time winner at Montlhéry, as its pilot? Or could the Delahaye firm, which had dazzled of late with its revolutionary 135, prove that the old French house was indeed a renewed force in motor racing? The presence of René Dreyfus and that American spitfire Lucy Schell at the autodrome showed that they were serious indeed.

Such was the drumbeat of questions, stirred keenly by frequent newspaper dispatches. With the French on their annual August holidays, there was plenty of time for predictions. "In bars and cafés, on beaches and golf courses up and down the country, it was a matter of fierce discussion," chronicled one writer. "[The Million] was a gloriously newsworthy, complex, emotional *mélange*—part entertainment, part chauvinism, part Russian roulette, as irresistible to the press as it was to the public."

Outside France, others were interested as well. The Germans, notably engineers from Mercedes, came to Montlhéry to spy on the potential prizewinners since they might have to face them in the upcoming Grand Prix season.

By August 10, François had finished the new engine and installed it in the car René had been testing. "The engine turned like a siren," René said admiringly: it was strong, powerful, and consistent, now enabling the Delahaye 145 to approach a maximum speed of 140 mph.

Together, René and François worked on optimizing its performance at Montlhéry.

At first, François disdained any of the driver's insights into "his car." What could René know about the technical side of things? He should just drive and report on how the car handled various sections of the course. Then François would decide what was best to do.

This did not sit well with René. In his time driving for Maserati and Bugatti, he had learned a great deal about the mechanics of a car. This understanding informed how far—and how hard—he could push an engine, gearbox, brakes, or suspension system. Such knowledge had often saved him from catastrophic failure—and allowed him to gain advantage over his competitors. René refused to cast aside these insights just because François disregarded them.

One afternoon when René returned to the pits, the situation came to a head. François criticized his timidity on the Ascari bend, the long left-hand turn named after the fiery Italian champion, Antonio Ascari, who was killed while navigating it during the 1925 French Grand Prix.

René took affront at the accusation that fear was holding him back. That had nothing to do with it. He argued that the Delahaye was incapable of handling the bend at the speed François wanted. Its tail would fling off the outside edge, just as Ascari's Alfa Romeo had done.

François disagreed. "It's very simple," René said, rising from the front seat. "Take the car, get in, and you'll see."

Certain in his belief, François climbed into the Delahaye. René handed over his goggles, and the engineer was quickly away. He returned from a first lap of the circuit. Unsuccessful. He tried again. Failed. One more time, and he came back. "I can't reach that speed on the Ascari bend," François admitted. "It's not possible."

René suggested a few changes, including lightening the car, to make the turn easier. Physics, not nerves, was clearly the issue.

"You know your business, I know mine . . . but I'm going to try," François said.

When François returned the Delahaye with the alterations made, René rounded the bend faster than ever before. From that day forward, François understood that there were some things that testbeds and calculations neglected to quantify. He needed to trust what René felt when he drove, whether it was a change needed in the gear ratio, carburation, braking, or suspension. With each day of tinkering, they wrested faster laps from the car, improvements measured in half seconds — and often less.

The sinuous circuit allowed these trimmings only begrudgingly. Montlhéry was very different from AVUS or Tripoli, courses characterized by long straights and easy bends that allowed drivers to go flat out and devour whole sections of the course. At Montlhéry, the many hairpins, sharp bends, right-angle turns, and plunging dips chewed up time. Some corners had to be taken almost at a halt. Powerful brakes were essential; top gear was limited to a few straights; downshifting was a necessary habit.

The design of the course was not the only problem. Its 12.5 kilometers of track and road were a dozen years old and reliant on a government that had little money for repairs. Protruding seams and unlevel sec-

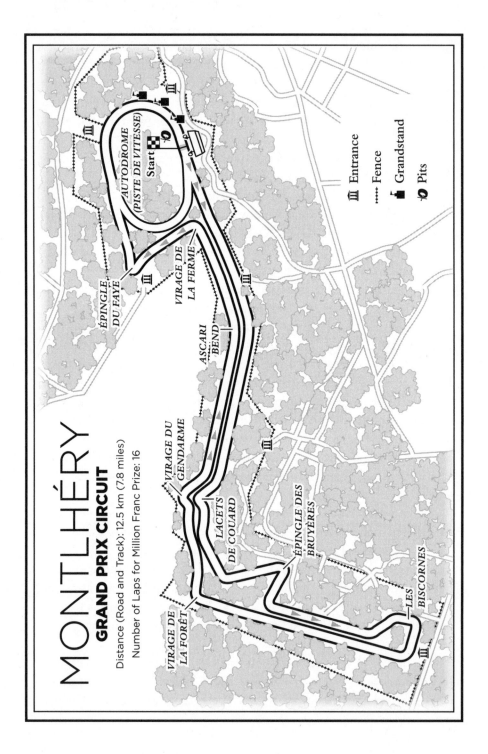

MONTLHÉRY
GRAND PRIX CIRCUIT

Distance (Road and Track): 12.5 km (7.8 miles)
Number of Laps for Million Franc Prize: 16

AUTODROME (PISTE DE VITESSE)

Start

ÉPINGLE DU FAYE

VIRAGE DE LA FERME

ASCARI BEND

VIRAGE DU GENDARME

LACETS DE COUARD

ÉPINGLE DES BRUYÈRES

VIRAGE DE LA FORÊT

LES BISCORNES

Ⅲ Entrance
······ Fence
🏛 Grandstand
🛢 Pits

tions bedeviled portions of the banked concrete oval, and the tarred road surface was mottled and cracked its entire length. In addition, gale-force winds often swept across the plateau on which the course was built. Although the summer months were usually calm, there were many days when gusts hit the cars like a wall.

The Écurie Bleue team understood the challenges. Every day, and often into the night, they worked to overcome them, as the August 31 deadline loomed ahead. René brought down his best lap time to five minutes, ten seconds (or an average of 145.2 kph over the sixteen laps). This was a dramatic reduction of almost thirty seconds from his fastest lap at the recent French Grand Prix, but it was still too slow to win the prize. Over sixteen laps, he needed an average of five minutes, seven seconds, and given the standing start, he had to be even faster when he got going — to recover time lost in the first lap.

Lucy knew that her team had more preparation to do. The Million Franc prize went to the team with the fastest time over 146.5 kph before the deadline, not the first one to achieve that speed.

Bugatti announced that it would make its attempt on Thursday, August 12. Mechanical trouble forced them to postpone by a few days, but on the Saturday, August 14, Jean Bugatti brought his firm's new formula car — a single-seater Type 59 fitted with a 4.5-liter engine — to Montlhéry.

Within forty-eight hours, he planned on attacking the Million. He had already summoned his champion, Jean-Pierre Wimille, back to Paris for the attempt. The driver's "boldness and mastery," the newspapers lauded, were sure to be proved again.

In the south of France, at his family summer home on the Mediterranean, Jean-Pierre readied to make the journey north. With him was his fellow driver Comte George Raphaël Béthenod de Montbressieux (nickname "Raph"), who was recovering after an accident at Le Mans. One more coffee and another cigarette and then the two friends would be off.

At the same moment, Ciafredo Beltrando, a thirty-two-year-old trucker from Aix-en-Provence, bicycled in his blue overalls to the lot where his employers kept their wine tankers. They were colossal machines, slow and lumbering as sea turtles.

It was a lovely day in Provence, the rolling hills cast in hues of yellow and green worthy of a painting. Jean-Pierre was driving a Bugatti grand tourer and making steady progress on Route 7, the national highway that led to the capital. Raph half dozed in the seat next to him.

Beltrando slowed his diesel-engine giant to a stop on a small road that crossed the highway, fifteen miles from his home. He was on schedule to reach the cooperative vineyard to fill up his heavy tanker — just another pickup in the workweek for him. Nothing special. After looking left, then right, and seeing nothing, he eased across the highway in first gear. There was a blind spot at the crossroads, which was set in a dip between the hills, but on a lazy August weekend there was seldom much traffic anyway.

At the exact moment that Beltrando was crawling across the highway in his tanker, Jean-Pierre came over the hill at 70 mph. He knew almost instantly that a collision was unavoidable. There was no allowance on either side of the narrow band of highway. His best chance was to jam on the brakes and try to hit the side of the tanker at the place where the greatest amount of steel separated him and Raph from it: head-on.

He shouted at his friend to brace himself as he hit the brakes. The tires hissed spitefully on the pavement, the back of the Bugatti fishtailed, and then they T-boned the tanker with a shuddering crash.

The two men were thrown forward against the dashboard and windshield but somehow remained inside the car. Such was the force of the collision that the tanker was forced off the road into a ditch thirty feet away.

Beltrando was startled, but unhurt. The two race car drivers were in bad shape. By the time they recovered consciousness, a crowd had gathered at the crash site. Several helped drag Jean-Pierre and Raph from their demolished Bugatti. Their faces were bleeding heavily, and their chests felt crushed. They were hurried to the closest town, where a doctor stitched and bandaged their immediate wounds, mainly some jagged cuts to their faces. He was unsure whether they had internal injuries, but they had undoubtedly suffered bad concussions. An ambulance transported them to a hospital in Marseille.

News of the accident reached Montlhéry early that afternoon. The shock was severe, inside both the Bugatti and Delahaye camps. Phone calls with the hospital reassured them that the two drivers would make a full recovery, but that they needed to be kept under observation.

Quickly, Jean Bugatti's thoughts turned back to the Million. His star driver was unfit to drive for at least a week. Team captain Robert Benoist might have to make the attempt instead. The hawk-faced forty-two-year-old had won four Grands Prix in a single season before Jean-Pierre was even out of school, but time had slowed his reflexes and dampened his fire.

Now little more than a fortnight away, the deadline would wait for no one. "Bugatti will have to fly," Jean told the press gallery. Printed alongside his words, newspapers across France carried photographs of the two hospital patients, wrapped in so many bandages that they looked like mummies. "Grievous Wounds" headlined one article. Despite his prognosis, Jean-Pierre told reporters that he thought he could attempt the Million soon.

On August 18, René was making a very fast practice run, and Delahaye looked like they might be ready for their own attempt, when the gearbox broke from his pushing the 145 too hard. They would be delayed from making any prize run for at least several days.

Jean Bugatti thought his injured driver might recover before then, but the Marseille doctors advised that they needed more time to monitor their patient because of his severe headaches. At the earliest, Bugatti now figured, his driver would be ready only a few days before the deadline. If there was poor weather or any mechanical trouble, any chances at the Million would disappear. He also had to consider that Jean-Pierre might be incapable of the physical effort required to race for 200 kilometers around the Montlhéry track.

Benoist prepared to go instead. In his first attempt, labeled for the press as a practice trial, he quit after a dozen laps because of strong winds. He did not run a lap faster than five minutes, nine seconds, and told reporters that Wimille could have easily bested his time. "Ah, if I was ten years younger," he added regretfully.

On August 23, the weather cleared and the winds calmed. Jean Bugatti was confident in his old champion and notified the press that the attempt would take place. They came with photographers and cameramen to record the prize run.

Benoist was very slow from the standing start. "Hésitant," *L'Auto* described it kindly. He finished the first lap in five minutes, twenty-six seconds, setting an average speed of 136 kph. This was nineteen seconds slower than the mean time per lap he needed to maintain for the prize, and it was a tough deficit from which to claw back.

In the second lap, he was again too slow—five minutes, ten seconds—and his time debt deepened. The Bugatti camp deflated. Benoist really had to get moving now. He gradually improved his times, handling the many turns and bends with inveterate skill. The Bugatti ran flawlessly. He finished in one hour, 22 minutes, and 3.9 seconds—nine seconds over the baseline target, a lifetime when it came to the Million bid.

In the pits, reporters asked Benoist if he would try again. "It's possible," he said before admitting that he hoped Wimille would recover soon. "He is faster than me."

Newspapers reported that the next attempt would probably be coming from René Dreyfus, who continued to practice every day on the road course. SEFAC was nowhere to be seen, nor Talbot. This was a contest between Bugatti and Delahaye now, and some suggested that the target was an impossible goal for both teams.

Across the country, headlines chronicled every move at Montlhéry. The spotlight on "Le Drame du Million" was reaching a blinding intensity, and the hope bloomed that France might actually field a worthy Grand Prix car for 1938. Furthermore, the prize was a welcome distraction from the otherwise consuming fears of impending conflict with Germany.

A gaggle of editorials chided the French government for its laxity in preparing the army for what looked like the inevitable. As one sarcastically noted, they were doing nothing to ready for motorized warfare. "All other armies were falling for the seductions of speed, for the promising charms of brilliant cavalcades [of armor] . . . It took a lot of courage to resist the contagion." The same was said for a modern French air force. One could not wholly blame those in power. Public opinion, a French historian wrote, could be summed up in four plain words: "Above all, no war!"

While the public spectacle of the "Million Franc" unfolded, Rudolf Uhlenhaut and his band of engineers and mechanics labored on their new formula car behind the high walls of Untertürkheim. No expense was spared. That year alone, the racing department cost Daimler-Benz

4.4 million Reichsmarks. The company's chief executive, Wilhelm Kissel, said that he supported the team out of the "national interest." This belied how the Nazi government rewarded the company in myriad other ways to compensate for their money-losing Grand Prix operation that employed hundreds and, after each race, disassembled and reconstructed every car.

Tax relief, busted trade unions, and highway construction—all had contributed to a recovery at Daimler-Benz that made it the envy of every other automobile company. Gross profits rose yearly, and production and investment returns were skyrocketing. The massive rearmament program played a critical role in the company's success. Daimler-Benz factories were churning out military equipment. With board member Jakob Werlin able to intervene directly with Adolf Hitler, the company won numerous large-scale government contracts. If—and very likely when—war broke out, the company's directors believed that its fortunes would only improve.

To maintain favor over Auto Union with the Reich government, Kissel knew he had to dominate the new Grand Prix formula as they had the previous one. They had begun planning many months before. Uhlenhaut led the effort, but the company also enlisted Ferdinand Porsche, who had finished his Auto Union contract, to consult with Mercedes on any new designs.

Given their long experience, there was little doubt at Untertürkheim as to which side of the formula—three-liter supercharged versus 4.5-liter unsupercharged—they would fall. At a meeting on March 23, 1937, they had decided on a V12 blown engine. The engineers proposed a chassis design based on their 1937 race car, but with the drive shaft angled slightly away from the center line, allowing space for the driver to sit even lower to the ground. Wind-tunnel tests on prototype bodies also led to improvements in the car's aerodynamics. After more development, these ideas were transformed into finalized plans, and the Daimler-Benz board green-lit production of a minimum of fifteen of the new model: the W154.

In August, while Mercedes engineers and mechanics built their race car, Neubauer sent his staff to keep an eye on Montlhéry. They noted the speeds the Bugatti and Delahaye formula cars could take in corners and their acceleration down straights. Neubauer was unconcerned that the French would prove any competition once the new season began, but it was better to be sure.

Monsieur Charles and Jean François thought they were fully prepared for a Delahaye attempt at the Million. Never patient, Lucy would have tried already. It was René who remained unsure.

On Thursday, August 26, he spent hours at Montlhéry, practicing. The gearbox's breaking had been a fluke, and it was a simple fix. The 145 was running very well. To increase its speed, the mechanics had lightened its weight by hollowing out any parts — including the chassis — that did not affect its structural integrity. Every ounce mattered. They even punched holes in the gas, brake, and clutch pedals.

At this point, there was little else they could do to the car. Responsibility for the prize now fell on René, and he knew it. Competing for the Million would be very different from a normal race. He would be alone on the track. There would be nobody behind pushing him, nobody ahead to pursue. There would be no opportunity to capitalize on an opponent's mistake, or to get a sense of the pace from the rest of the field. There could be no blaming interference by another car. He would be driving against the clock alone.

In many ways, René thought, the prize run was very similar to a hill climb. Everything depended on a driver's ability to maintain discipline and push the car to its limits. If he wavered, the clock would not compromise; it would not care.

He had lost count of how many times he had lapped the course looking for ways to shave off seconds. Many hundreds — probably more. The cockpit felt as comfortable and well-worn to him now as an old armchair. He knew every bend and hairpin on the circuit, every hollow and bump in the surface, every marker — whether a tree, shrub, gate, fence line, or undulation in the hill — that would signal to him where he would soon be. He knew how fast to take each stretch of the course, where to brake, when to shift, how wide to drift into a turn, at what point to exit it. He knew the angle to enter the autodrome from the road, the height to follow along its banked curve, and the gear to use when shooting through the opening back onto the road circuit. He felt capable of navigating the course at night without lights.

Still, he felt unready. The offside rear wheel rose slightly from the pavement when he accelerated too much into some turns. This was costing him time. To avoid this he needed to know exactly the point at which to stop accelerating in each turn.

Above all, he had yet to consistently beat the threshold of five minutes, seven seconds.

"You must try tomorrow," Weiffenbach urged. The weather looked fine. There was very little wind expected. Also, Jean-Pierre was flying back from Marseille that same day—in fighting shape, or so he was telling reporters. The Bugatti driver might make an attempt as early as Saturday. If he broke the 146.5-kph average first, the pressure on René would only increase. They might even find themselves in a pitched battle to establish the fastest time before the August 31 deadline.

"Dreyfus, I trust you," Monsieur Charles said. "You will go, and you will succeed."

As much as he respected Weiffenbach's opinion, René wanted more practice. Another few tenths of a second per lap might prove essential. He thought that his standing start could be better. Another day at least was needed. He did not want to try and fail. He wanted to be sure he could do it when he set off. Caution was overriding his newly rediscovered combativeness.

"No, I can't do it," he said. "I don't have enough leeway. I'm losing too much in the early laps."

"No, you're not."

"Yes, I am."

Monsieur Charles looked at him. He had always told René that the decision to go rested with his driver and with him alone. "All right. *Bonne nuit,*" he said.

That night, back at their apartment on the western edge of the city, René and Chou-Chou had dinner together, then took their hound, Minka Stuck, for a walk. The little dog was a gift from Hans and Paula Stuck—the couples had become close over the years, not least because of the persecution both faced because of anti-Semitism. Their friendship was representative of the complicated nature of relationships between Grand Prix drivers. After all, Stuck had personally campaigned Hitler to support racing—and record attempts—while at the same time suffering from the prejudices he stoked. These pressures weighed heavily.

René and Chou-Chou watched the sky darken over Bois de Boulogne. In the late summer, the birds sang well into the evening. Sometime after 10:00 p.m., they readied for bed.

Chou-Chou was already in her pajamas when their phone rang. She quickly answered it and spoke in whispers.

After a few moments, René asked, "Who is it?"

"Monsieur Charles," Chou-Chou said before drawing away again.

"What is it about?" René asked, now hovering.

"Oh, nothing at all."

"If it's about doing the Million tomorrow, I'm not going."

She waved him away and talked some more to the Delahaye boss. Finally, she hung up and returned to the bedroom.

"What did he want?" René asked, exasperated.

"Oh, nothing!" Chou-Chou said. "He just wanted to know your impressions from your practice earlier. That's all."

Exhausted, René went to bed. His plan for the next day was to start with a nice long lie-in, followed by an afternoon of practice at the track.

In what felt like mere moments after he had fallen asleep, the alarm clock on his bedside table clattered. He squinted at it. It was only 5:30 a.m. He turned to Chou-Chou and grumbled, "There are no tests this morning." He could not understand why she had set the alarm.

"It's all ready," she said, rising from bed. "Come on and get up, René. You're trying the Million today at ten o'clock."

René shook his head. No. No way. He needed more practice.

Chou-Chou told him that he would never be completely prepared but that he would have to try anyway. René told her that she was wrong. He would reach the point when he was sure he could do it. He just needed more time. Monsieur Charles and Lucy believed differently, Chou-Chou said. They were both convinced that he must shoot for the prize today. Hence the phone call late the night before. They did not want to give him the opportunity to back out.

"No!" René blustered.

"Shut up, calm down, and get dressed," Chou-Chou snapped. The timekeepers were on their way to the track. The press had been informed. As would everybody in Paris as soon as the morning's papers hit the newsstands, since each would carry the news of his scheduled attempt.

His hand had been forced. René steeled himself to go.

11

The Duel

RENÉ DROVE OUT of Paris in his black Delahaye coupé, Chou-Chou sitting beside him, Minka on the back seat. René was wearing a jacket and slacks, his wife a striped blue blouse and flannel skirt. They looked like they might have been heading out to the countryside for the weekend in the elegantly bodied car with its swooping art-deco lines.

Instead, René steered toward Montlhéry. His face wore a grim look, and there was a stilted silence in the car. Although resolved to try for the Million Franc prize, René was deeply cross over the deception by his wife and race team. He suspected that his anger had been stirred on purpose by Lucy and Monsieur Charles to make him more aggressive during the run. The thought that he could be played so easily only upset him further.

Once off the highway, they followed a rough road toward the plateau where Montlhéry stood. The top of its grandstand eventually appeared on the horizon. At the entrance, they passed some buildings where the track's administration staff worked. Then the slender concrete columns that supported the autodrome's banked bowl came into view. An *Autocar* writer described them as "branchless and petrified trees." To stand under the concrete bowl while cars raced overhead was akin to being in an earthquake.

René parked and then headed to a dressing room underneath the autodrome to change into his white racing suit. When he emerged out of the tunnel, he found a crowd of journalists and timekeepers on the track. Still furious, he avoided greeting Monsieur Charles or Lucy.

The Delahaye 145 was ready in the pits. Its body remained a bare aluminum shell, ugly as ever, with no Delahaye or Écurie Bleue markers. Its engine was already warmed up by the time René arrived. François and his crew had worked overnight to ensure that everything was tip-top with the car, but he wanted René to take a few laps of the course first. René took his time. It was a gray morning. The pavement was dry. Occasional gusts of wind swept across the plateau, and he knew that any headwind would slow him down. If given the choice, he would have delayed the run. But the decision was no longer his to make. Everybody was on hand and eager that he go.

After a break in the pits, René performed his ritual check of the triple knot on his shoelaces, then climbed back into the cockpit. Everyone was in position, including Chou-Chou, in the timekeeper's stand by the starting line. She would relay his lap times by telephone to a Delahaye mechanic stationed in a water tower a kilometer into the course. That way, René would know almost immediately how he had done on the previous lap and whether he needed to push faster.

He rolled the car toward the starting line. He was very anxious about his chances, particularly because of the wind. He had nothing to say to Lucy or to anybody else on his team. Monsieur Charles came up beside the 145. "You have our confidence," he said.

René barely nodded in reply.

Leaning against the wall of the track was Robert Benoist. In the grandstands, also taking note of the competition, was Prince Wilhelm von Urach, a manager for the Mercedes race team. With binoculars around his neck, he made no effort to hide his presence. A herd of French and foreign journalists, and even a cameraman, were on hand to record René's attempt as well.

René placed his hand on the gearshift and readied his feet on the clutch and accelerator. The twelve-cylinder engine reverberated through to his grip on the steering wheel. The weight of everybody's expectations fell heavy on him. So too did his desire to prove that he had reawakened the fearless instinct that saw him first claim Grand Prix victory. That he deserved to be fielded against the best drivers in the world, regardless of politics.

Sixteen laps. Two hundred kilometers. In one hour, twenty-one minutes, and fifty-four seconds or less. Every second counted, particularly in the first lap, which was begun from a standing position.

The starter raised the flag. René revved the engine, its guttural rasp breaking through the quiet morning air. His blood felt frozen in his veins.

At 10:00 a.m. sharp, the flag dropped, and the Delahaye 145 leaped forward, its tires smoking on the concrete track.

René swept past the grandstand and dove off the bowl through a narrow opening onto the tarred road that led from the autodrome. It hadn't been the quickest start, and his arms felt deadened by nerves. He shifted into top gear to take the downhill two-kilometer straight at top speed. To his left was a scattering of trees and heather. To his right, a slender ribbon of grass and wildflowers separated this outbound road from the one he would return along to the opposite side of the autodrome.

At the end of the straight, which he took at almost 210 kph (130 mph), the road drifted to the right before reaching the Lacets de Couard. The easy part of the course was over. He rapidly shifted gears through three bends, then the road pitched at a precipitous downgrade into Épingle des Bruyères, the hairpin that had to be taken at a creeping pace. Shortly after that, he crossed the four-kilometer marker and braked as he entered another turn. He could only briefly hit fourth gear in the straight that followed before Les Biscornes, the westernmost point of the circuit. When he emerged from its series of right-hand bends — it resembled a square — he was halfway through the course.

His driving was not effortless. He worried that he had lost too much time. But there was no way yet to tell.

Another kilometer straight followed. It had an uphill grade and could have been taken at top speed throughout except that it was interrupted by a troublesome dip that would have shot his car skyward had he tried. He pushed as fast as he could. Gusts of wind struck the side of his Delahaye, bucking it slightly off his line.

Just past the seven-kilometer marker, the road rose steeply again into the Virage de la Forêt, a right-angle corner by a stand of trees. Shifting out of it, René then darted through two gentle bends, making sure to keep his tail from sliding out sideways because of his speed. The course was now generally at an even grade, but his arms and legs got little rest as he was suddenly into another sharp corner. When he straightened out of it, René followed the road that ran parallel to the outward-bound stretch of the circuit. Again he hit top speed, a haze of dust behind him.

After the ninth kilometer, he made the long left-hand bend where Ascari had flipped over and died. René survived the sweep into the straight that followed. Ahead rose the columns of the banked autodrome. He reached the sharp Virage de la Ferme, then tracked the edge of the bowl before braking at the Épingle du Faye, another hairpin.

René on the Montlhéry road circuit during the Million Franc run in August 1937

Desperate to know his time, René barreled through the slender opening back onto the wide concrete surface of the autodrome. He was now opposite the straight on which he had started and was almost finished with the first lap.

He took the eastern banking of the bowl very fast, centrifugal force keeping his tires on the concrete. Out of the bend, he dove into the straight by the grandstand, making sure to correct from any swerve caused by where the banking met the flat.

Chou-Chou clicked her stopwatch to register René's time. Onlookers cheered as he passed, their voices lost in the snarl from the 145 engine. One lap down. Fifteen to go.

The crowds tracked the Delahaye as it thundered out of the autodrome again. They would see very little of it, just hear the screech of its tires and the echoed pitch of its engine altering as René transitioned through the gears in rapid succession.

Soon after, at the water tower, René looked for his time on the blackboard.

Five minutes, 22.9 seconds.

Slow. Way too slow. "That's not enough. I won't do it," he told himself. He needed to average five minutes, seven seconds on each lap.

Even with the standing start, it was a crushingly bad time. To claim the first tranche of the Million fund for Bugatti, Wimille had run the same lap five seconds faster. René had a nearly sixteen-second deficit to recover from now.

He maneuvered through the road course, his arms and legs still feeling leaden. He needed to settle down. He needed to trust himself. The 145 was running well. He had to do better—much better.

He swung the wheel back and forth through the Lacets de Couard, braked hard at the Épingle des Bruyères, rounded out of Les Biscornes, and sped along the swift straight before the Virage de la Forêt. After another stretch of the course and two sharp turns, he returned to the autodrome, then slingshotted out of the banked bowl past the grandstands again.

Lap two done. He soon received his time at the water tower: five minutes, 10.2 seconds. Again too slow. He had done nothing at all to chisel at the deficit and had only added to it.

Nineteen seconds to make up now.

Discipline. Precision. He needed both to make the necessary average. If he failed to hit it on the third lap, his attempt might well be lost. He pushed faster into and out of each turn. Started to feel the course better. His reactions quickened.

Third lap: five minutes, seven seconds. On mark. Average speed: 146.6 kph. It was well won, but René had to improve and knock off one second, maybe two, per lap to climb out of the hole he had dug for himself. Over the next three laps, he managed to average the same: five minutes, seven seconds. It was not good enough. Ten rounds of the circuit remained. Now he had to reduce his times almost to five minutes, five seconds on each and every lap. He was already pushing at the edge of his ability — and at that of the Delahaye.

Eking out a couple of seconds over a 12.5-kilometer course was an immense challenge. René was already taking the best line into each hairpin. His speed in the straights, down the hills, and around the long bends approached the limit of what the car could bear and hold the road. The brakes and tires could only be strained so much. At best, he might shave tenths of seconds out of each lap.

Any error or misjudgment would cost dear time, but worse, it might launch him off the course or over the top edge of the autodrome. Drivers had died at Montlhéry in similar record attempts, which were no less dangerous than real races. Shortly after René left Maserati, one of its drivers, Amedeo Ruggeri, had lost control on the Montlhéry bowl during an attempt at the World Hour Record. His car somersaulted five times before coming to a standstill on the track, and he never regained consciousness. René's close friend Stanisław Czaykowski had catapulted off a similar autodrome during the Monza Grand Prix. René had to force away such thoughts — and those of his own crashes in the past. He needed to match his course discipline with boldness now.

In the seventh lap, the wind picked up. Any tenths of seconds stolen by the gusts René committed to recovering. He blazed around the course, eyes trained on the bumps, dips, and turns ahead. The stretches of grass on the roadside passed in brushstrokes of green. The slim windshield barely cut down the rush of air that buffeted his cheeks. As he cycled through the gears, the bark of his engine deafened him. His body shook as the Delahaye danced over the bumpy sections of the tarred road.

A hard left. Then a straight. Then a stomach-plummeting descent into the hairpin. No mercy on the brakes. Cut the next corner sharp — gain inches. Another straight. Left. Then a long, seemingly endless right bend. After coming around the sharp Épingle du Faye, René whipped into the autodrome. His field of vision down the hood tilted on its side as he drove up the bank. Sky to his left. Concrete to his right. Orientation became difficult, and he had the sensation of being a fly clinging to a wall. There was no rail or fence to stop him from soaring over the top, and every depression in the concrete made his car bounce. He stuck tight and then flung down into the straight.

His time on that lap: five minutes, 5.6 seconds. That was it. Much better. He had won almost 1.5 seconds from his deficit. Yet he had to coax even more speed out of every bend. After the eighth lap, the half-way point, he quickly learned that he had done just that, finishing in five minutes, five seconds flat. Now he was only fifteen seconds down from a Million win. His confidence was growing.

Over the next four laps — nine, ten, eleven, twelve — he fought the bucking and heaving of the wind to manage slightly better than an average of five minutes, five seconds through the course's dozen bends, eight corners, and two hairpin turns. He was in rhythm, feeling in perfect union with the Delahaye and their assault together on the course. They must keep to it. Every moment.

Fewer than six seconds remained in their deficit.

In the stands, Lucy and Laury watched the clock closely, willing René to go faster. Down by the pits, Monsieur Charles and François were doing the same. Their car and their driver needed to stay strong through the final four laps.

Not a few journalists with their own chronometers felt like Delahaye had a shot now.

René threaded through the thirteenth lap, sure that the constant braking and accelerating were wearing his rear tires thin. He knew this was a risk, and a Dunlop representative was stationed on the road course to alert him if he saw one of the rubber treads giving way to canvas and threatening a burst tire. Such an event would be cataclysmic to his attempt — and maybe even his life. So far, the Dunlop man had given no indication that anything was wrong.

Returning toward the autodrome in the thirteenth lap, René smelled something acrid in the air. Something burning. Ahead at the Épingle du Faye, smoke billowed across the hairpin. In a nearby field, a farmer was burning weeds. René should have slowed. The smoke obscured the contours of the road. But he could not lose the time.

He knew the turn by heart and decided to take it on faith. He pressed his foot on the gas and emerged out of the hairpin and its dense gray screen to shoot back into the autodrome and finish another lap. Five minutes, 5.3 seconds.

He was within striking distance of overcoming his time debt. Three laps to go. In the fourteenth, René blistered through the straights and turns and clocked his best lap yet: five minutes, 3.9 seconds. He was now less than half a second in the hole.

Everybody in the autodrome, most of all Lucy, who was watching the Delahaye speed past into its fifteenth lap, sensed the possibility of the Million.

René only needed to run slightly better than five minutes, seven seconds. He had done so in the past half-dozen laps. Journalists scribbled in their notebooks. The cameraman tried to keep his lens trained on the car as it disappeared through the narrow gates back onto the road course. Chou-Chou rapidly spoke into the phone to the water-tower relay to make sure her husband knew how well he was running. The Delahaye boss and his engineer prayed that the engine would hold.

René whipped down the first straight toward the Lacets de Couard, then down to the Épingle des Bruyères. The Dunlop man, Kessa, was down in the grass as the Delahaye clung around the hairpin, frantically waving his hands. He could see the warning strip of the white canvas underneath the rubber. The tire's tread was wearing off and it might burst at any moment.

Tire be damned. René continued. He could not hesitate or falter. He finished the fifteenth lap in five minutes, 4.3 seconds. Not only had he erased his debt, but he now had time in the bank. Given the threat of the tire bursting, he might have done well to ease off, but he refused to think that way. He urged himself to be braver — bolder.

When he returned to the hairpin on the next lap, he glanced at the Dunlop rep. Kessa had a look of horror on his face. René passed him quickly but could have sworn he saw the man make the sign of the cross. Onward. Onward. He rounded Les Biscornes, shot up the straight to the Virage de la Forêt, and coursed through the series of bends until the return toward the autodrome. He never slackened, crossing through the smoke that drifted across the Épingle du Faye like a man possessed. He swept around the bowl again and slung down past the checkered flag by the grandstand.

The timekeepers clocked five minutes, 4.5 seconds. It was a triumph. In the best — and hardest — drive of his life, René qualified for the Million with an average speed of 146.7 kph in a time of one hour, twenty-one minutes, 49.5 seconds—exactly 4.9 seconds under the limit. Écurie Bleue and Delahaye owned the prize if nobody bettered René's time by August 31.

When René came back around the autodrome, a crowd awaited him on the track. He rolled to a stop, pulled his goggles off his eyes, and lifted himself up onto the back of his seat. Dirt flecked his overalls, and soot covered most of his face, which only made his smile seem all the brighter. Monsieur Charles and François reached him first, congratulating him with bear hugs. The Schells, sons and all, gathered around the car to cheer him. Delahaye mechanics and staff leaped in joy. Finally, Chou-Chou arrived down from the timekeepers' stand to give him a kiss. As she sat on his lap, photographers snapped dozens of shots for the next day's front pages.

As the realization set in that he had done it, René stood away from the 145 and wept, in both joy and relief. Lucy Schell and the whole Delahaye crew joined him in his tears. The pop of champagne corks followed, flowers were draped around an exhausted driver, and the whole

René claims the prize, August 1937 (left to right: Charles Weiffenbach, Chou-Chou, René, and Jean François)

crowd—journalists and all—headed to La Potinière, the restaurant beside the autodrome, to celebrate.

A report came in that Jean-Pierre Wimille was flying into Paris that very moment and would make his bid for the Million before the deadline—perhaps as early as the next day. René tried to not allow the rumors to dampen his enjoyment of a well-earned drink.

"4.9 Seconds Are Worth 1 Million," the headlines announced the next day. The reports beneath carried a detailed, lap-by-lap breakdown of René's triumph. He was quoted far and wide, thanking the Delahaye team, its selfless mechanics, his "perfect" car, and his team boss Lucy Schell. "You understand why I don't yet shout victory," he said. "Tomorrow or Monday or Tuesday another competitor will take his chance, but what I can say is that the task of my adversaries will be as difficult as mine." He concluded with a simple statement that revealed how much he had changed since first joining Écurie Bleue. "To win the prize, it is necessary to take great risks."

Lucy hardly slept thinking about what might happen next. She was up early, reading the newspapers and listening, in her drawing room at home, to radio reports about the contest between Delahaye and Bugatti for the Million Franc prize.

Jean-Pierre had arrived at Montlhéry in the early evening of the day before to test-drive the straight-eight, 4.5-liter single-seater in which he would make his bid. Pale-faced, with a deep cut marring his upper lip, he wore a white bandage around his half-shaved head. But there wasn't the slightest hesitation in his step when he climbed into his Bugatti.

After a few laps, he came into the pits, complaining of some gremlins in the engine. The mechanics promised everything would be ready for a run on Sunday, August 29.

Lucy planned on having René and her 145 ready at the autodrome to run again if Bugatti bested their time. On her orders, François and his mechanics were already inspecting and overhauling the car back at the factory. They replaced the tires, changed the oil, switched out the plugs, flushed the filters, greased the suspension, modified the springs, and tightened every nut and bolt within reach. They also washed and shined its mottled aluminum body, truly an act of kindness for the ugly little machine.

Early on Sunday morning, after another restless night, Lucy drove to Montlhéry. René and his wife were there as well. In a bid to hold on to their luck, René wore the same overalls from his prize run, and Chou-Chou the same blue-striped shirt and skirt. Lucy noticed that Monsieur Charles and François had also stuck with the same suits and ties they had been wearing on that Friday.

Everybody was tense, especially René, who might have to go out and better his performance. They spent the day waiting for Jean-Pierre and the Bugatti team. Hour after hour passed. Their rival never showed up.

Later that evening, the press reported a statement from Jean Bugatti. "Tomorrow morning, at 8 o'clock a.m., an official attempt will be made."

Lucy returned to La Rairie to spend another night stewing over what the newspapers and radio announcers called "The Race for the Million."

Monday morning, August 30, and another day of waiting at Montlhéry began. A sea of reporters were on hand, as well as timekeepers and ACF officials to oversee the Bugatti attempt. Again, nobody from the firm appeared. There was no statement this time.

After sundown, Lucy and her team departed. One frustrated sports commentator wrote, "Since everything has an end, this one will be delivered Tuesday — at the twelve strokes of midnight."

On the third and guaranteed final day, Lucy arrived at Montlhéry early. Yet again everyone wore the same clothes, except Chou-Chou, who had finally had enough of the old racing superstition. René was a wreck. The constant waiting had frayed his nerves, and the lack of sleep had left dark bags under his eyes.

At 7:00 a.m., he ran a couple of test laps with the 145. It performed well again. The weather was clear and windless — it was an ideal day for any attempt. He spent the rest of the time in the pits, sitting in a chair, fretting.

They all tried to keep him relaxed, spending moments at his side, offering him a drink of water, a smoke, a mild aperitif. Nothing worked. One hour after the next ticked by, and there was no sign of Bugatti. If they didn't show, the August 31 deadline would see the Million awarded to Delahaye.

By the Fonds de Course rules, only a French manufacturer could accept the prize. Lucy knew that it had been her ambition—not to mention her money—that had seen their effort through. She did not court the attention of any reporters, despite their irksome habit of assigning leadership of Écurie Bleue to her husband. She had come to expect such slights, and no doubt they were part of what fueled her ambition. At least Monsieur Charles had been gentleman enough to promise that the million francs would be split evenly between her and his firm if they did win.

As lunchtime passed, Lucy could only guess at their competitor's intentions. "Were Bugatti and Wimille holding cards they had not yet played?" she asked herself. "Could they be conjuring up some sort of trick? Suppose the Bugatti turned up at 6 o'clock in the evening. It would not finish its run until 7:15 p.m. The Delahaye could hardly start another run before half-past—and the daylight would have begun to fade long before they could complete it."

Considering that Bugatti might well attempt this gambit on the deadline, she and Weiffenbach petitioned the fund's committee members to allow René to make his own follow-up run. Nothing in the rules forbade this, and the committee agreed that the 145 could start two minutes after Jean-Pierre began—if he began at all. They would share the course, spiking excitement that "The Race for the Million" would indeed be a race after all.

The potential for such an event saw motorsport enthusiasts flow into Montlhéry. Newspaper reporters had hoped to post a winner by their evening editions, but the presses had to run with the question unanswered: "Delahaye: Temporary Keeper of the Million—Will They Be Beaten by Bugatti Challenger?"

At 4:00 p.m., Lucy spotted a Bugatti truck carrying a single-seater car emerge from the tunnel. The car was unloaded onto the track by a gaggle of blue-coated mechanics. Jean-Pierre showed soon after in a racing suit of the same color. The veteran Benoist hung close to him, whispering words of advice and calm.

They had come at last.

The two teams greeted each other. René shook hands with a tight-jawed Jean-Pierre, the mood akin to two boxers bumping gloves before the first ring of the bell.

Jean-Pierre climbed into his car and took off around the road circuit for a test run. Minutes passed, many more than the lap should have taken, before he returned to the autodrome. The single-seater was making an awful racket. Something was wrong.

Lucy and the rest of her team watched on tenterhooks as a squad of ten Bugatti mechanics circled the car. Word came that the rear axle was broken. It was a significant problem, surely unresolvable before sundown and the midnight deadline. Odds were that the Million was now firmly in hand. René might not have to run again. He allowed himself a slight smile.

Remarkably, after over an hour of clanging, welding sparks, and feverish activity, Jean Bugatti declared the car fixed and ready for the attempt.

With the sun falling toward the horizon, René slotted his body into the 145 and straightened his white linen cap. He had already earned the Million Franc prize, and he was not about to relinquish it to anybody, let alone Jean-Pierre. René was determined to drive even faster than before. The many hours spent over the many days at Montlhéry, particularly the ones just past, had left him anxious but also charged up for a fight. He felt a distinct fire in his veins.

Jean-Pierre was ready at the start. The sky was tinged pale pink, and lights were already glowing around the road circuit and autodrome. It would be almost dark by the time he and René finished their runs, which only made their attempt even more dangerous.

At 6:43 p.m., the Bugatti launched away from the autodrome, its engine screaming into high gear as Jean-Pierre took the straight away from the track.

René sat in the 145. Thirty seconds passed. A minute. Thirty seconds more. The countdown began: ten-nine-eight-seven — he prepared to go, feet steady on the pedals — six-five-four-three-two-one . . . The chase was on.

Jean-Pierre would already be taking the square turn at Les Biscornes. Mastering his nerves, René sped after him as if he needed to overtake his rival rather than simply beat his time over sixteen laps. The kilometers fell away like he was being slung around the road circuit on a rail track. His line into every bend and turn was pure and fast, his reflexes instantaneous. He rounded the autodrome to finish the first lap, conscious that he had never sighted Jean-Pierre, nor heard the Bugatti engine over his own.

He streaked past the crowded grandstands. The pits were choked with both teams of mechanics. Journalists and radio announcers (and much of France and beyond beside their wireless sets) took note of his time long before he knew it himself. Five minutes, 19.6 seconds. Passing the water tower on his second lap, René eyed the board held aloft by the Delahaye mechanic. He was 3.3 seconds faster than the standing start in his first Million run but less than half a second better than the chalked time of his rival.

René accelerated. Minutes later, he turned the 145 around the banked bowl and passed the pits. There, surrounded by mechanics, he spotted the Bugatti car, motionless. Jean-Pierre was waving his arms, urging the mechanics to hurry. Benoist stared at his chronometer, a sober look on his face, as the Delahaye tore past. The Bugatti was firing on only seven cylinders — one of its spark plugs was mucked up. Jean Bugatti insisted that as soon as it was fixed they would commence another attempt.

René knew only that he planned on finishing what he had started. He had clocked five minutes, 8.1 seconds in his second lap, a 2.1-second improvement on four days before. Seized by the pure joy of going faster and faster, he finished the third lap at an even better pace. If he continued at this rate, he might cut half a minute off his time.

As René sped into his fourth lap, the Bugatti was ready again.

"It's too late," Jean-Pierre said, peering into the sky.

"Go!" Bugatti insisted.

At 7:02 p.m., the single-seater sped away. Now Jean-Pierre was the one chasing René. The two whipped around the road circuit.

Coming back toward the autodrome on his fifth lap, René caught the acrid scent of burning oil. He knew it must be from the Bugatti, which would have passed seconds before on the parallel outbound road. Something must be wrong. He advanced into the sixth lap unaware that Jean-Pierre had pulled into the pits, smoke billowing from his exhaust. Oil was leaking from the engine, and there was no more time left in which to repair it. Their attempt was over.

René continued his race, tires screeching in the hairpins, swooping around every bend. Only when he returned to the autodrome and received the signal to slow down did he realize it was over. He came around toward the straight by the grandstand and saw the trackside swelled with people. He finished the lap—five minutes, seven seconds—and pulled up quickly to a stop.

Again Lucy and the Delahaye crew surrounded him. Again Chou-Chou wrapped herself around his neck. Again there were cheers and smiles. Jean-Pierre threaded a path through the melee to shake his hand. "Bravo! The best man has won today," he said. "Next time I will have my revenge."

The triumph of "Le Duel Delahaye-Bugatti" dominated front-page news the next day. Praise in speeches and editorials abounded for René, who had "driven like a god." The entire Delahaye team won their share of the congratulations as well. The Million Franc prize, one chronicler noted, was the "jewel in the dazzling crown" of the rejuvenated rue du Banquier firm. "It is a unique diamond that shines a thousand lights." Weiffenbach was heralded as a living testament to tenacity, while François was no less than a designer extraordinaire.

Again, seldom was mention made of Lucy Schell, despite efforts by René to highlight her role as the "creator of Écurie Bleue." He stated that her contribution to their success was unparalleled. She knew as much in her heart, but the lack of recognition burned, even if it was to be expected.

Over the following weeks, René rested at home and received a stream of congratulatory letters. One was from the head of the automobile club responsible for staging the Pau Grand Prix. "Will we count you among the lucky winners at our race? We hope so."

Some of the letters referred to René as a millionaire, even though his share of the prize was only a quarter of the total sum. (Lucy shared her half with him, fifty-fifty.) Still, 250,000 francs was almost three times what he would have earned winning the French Grand Prix. The money was insignificant compared to the confidence the win had given him. In his pursuit of the Million, René had proved to himself at last that he had the necessary killer instinct to be one of the greats.

France had a new national hero. To those who knew of the Dreyfus Affair, it must have seemed odd to see that name in the headlines and to realize that a race car driver of the same name had brought the country together.

Lucy was only getting started with him. She wanted René—and

Delahaye—to prepare for the first Grand Prix race of the 1938 season. Victory there would make a statement for the rest of the season. It was one thing to win a time trial, but quite another to win an actual race—and in doing so, beat the Germans in the arena they had ruled for years now.

Promotional ad by Delahaye after the Million Franc Race

12

"One of Us Will Die"

ONE MORNING IN October 1937, the French president, Albert Lebrun, arrived at the Grand Palais des Champs-Élysées in a black limousine. A band struck up the "Marseillaise" as he stepped from the back seat of the car for the opening of the Salon de L'Automobile. His presence at the show was purely ceremonial, and considering the impatience with which Lebrun toured each stand, it was clear that he saw it as something of a chore.

In the center of the vaulted glass hall stood the Delahaye 145, now painted a striking blue, on a revolving dais. The hood of the race car had been removed, revealing the polished V12 engine in all its glory. The winner of the Million attracted some attention from Lebrun, but he quickly moved away to the next exhibit, then the next: Citroën, Renault, Delage, Bugatti, Talbot, Peugeot.

In total, he spent a half hour at the Grand Palais. His lack of attention deflated the show's participants' hopes that the president might be a champion for the country's automobile industry. After another dismal year when the secondhand trade outperformed sales of new models, they desperately needed the boost.

Delahaye was one of the few thriving French manufacturers exhibiting at the Salon, and Charles Weiffenbach was an authoritative voice there, thanks to the company's success at the Million. He advocated the same kind of state support that Hitler had granted Germany's automotive firms, whose sales were booming. "Our French automotive industry," Weiffenbach stated, "is now the poor relative, after so many years its most brilliant representative." Further, he noted, when a Mercedes or Auto Union car won a big race, nobody cared about the brand. It was a "German car," and the benefit to the country's reputation worldwide was clear.

The speakers at the ACF's annual dinner after the show struck a similar chord. Among the honored guests again were René Dreyfus and Rudi Caracciola. René accepted an award for his Million win, while Rudi was celebrated as the overall European Champion.

Several Mercedes officials, including Prince Wilhelm von Urach, reported back to Untertürkheim after their visit. "The Paris Show makes a very unfavorable impression," said one. "There was a noticeable stagnation in the development of French cars," commented another. They noted the prominence of the Delahaye 145 at the show but praised neither its design nor its performance in winning the French prize. "Not too impressed" was their takeaway: The supercharged three-liter Mercedes would far outpower the Delahaye, and it would simply not be able to compete.

Punctuating the end of the Salon was a newspaper editorial from Charles Faroux. He wrote not of cars or races but of war. In his mind, France was in "imminent danger" from Germany, whose military expenditures and army ranks swelled every year. He cited a litany of statistics: the equivalent of 300 billion francs spent on rearmament, or 75 percent of the total Reich budget; 1 million men in the German army; and 1,000 bombers able to destroy Paris in a "single night." Faroux concluded that Hitler was deceiving the world with his peace proffers.

"France, do you understand?" Faroux pleaded.

At the airport outside Frankfurt, near a Zeppelin airship hangar, Rudi stepped up onto a wooden platform to reach the cockpit of the streamlined Mercedes. Built to break records, the car was based on the W125 chassis but fitted with a V12, 5.6-liter supercharged engine that delivered enormous amounts of power, more than any car ever built by the company. The body, tested for resistance in a wind tunnel, wrapped around the whole vehicle, even its wheels, and looked like a silver tortoiseshell flattened low to the ground.

When Rudi slid into the driver's seat, only the top of his head could be seen by the crowd of onlookers, including schoolchildren, who stood outside the boundary fence for the launch of Reich Rekordwoche (Reich Record Week) on October 25.

Since motor cars were invented, speed records had seized the public imagination. In 1898, Count Gaston de Chasseloup-Laubat, a Frenchman, piloted a Jeantaud electric car at 39 mph to establish the world's first land-speed record. In 1935, British driver Malcolm Campbell set a new mark at 301 mph on the Bonneville Salt Flats in Utah with his rectangular-shaped *Blue Bird*. Campbell's car, if one could call *Blue Bird* a car, was a twenty-seven-foot-long behemoth with a 36.7-liter aeroengine, but for the most part the vehicles that claimed records were not very different from their everyday counterparts.

There were many records to be won: fastest speeds over distances from one kilometer to 180,000 miles, from both a standing and a flying start; or the greatest distance traveled over a certain time, from one hour to 133 days. Overall world records had no limit on engine size, but world class records were based on capacity.

From the moment he attained power, Adolf Hitler set about having Germany break automobile records, and he mandated investments in both Mercedes and Auto Union to make it happen. Hans Stuck was the lead driver in these efforts. As early as 1934, he set several world class records at AVUS for Auto Union. Rudi Caracciola followed with numerous records of his own for Mercedes. In this way, the feud between the two firms, most visible between Rudi and Bernd Rosemeyer, carried on beyond the Grand Prix.

With its long stretches of ruler-straight roads, the new autobahn provided a perfect venue for challenging speed records when traffic was shut down. The first bet was made in 1936, between Hitler and Fritz Todt, the "Inspector General of German Roadways," on whether the journey between Berlin and Munich could be run in less than three hours. Todt won the wager after Stuck managed the 260 miles in two hours and seventeen minutes. Hermann Göring, the Reich aviation minister, tracked the driver's progress overhead in a Junkers Ju 52.

The following year, shortly before leaving for the Vanderbilt Cup, Bernd stole several records from Rudi, including the Flying Start Kilometer Record on the autobahn, in which he averaged a speed of 242 mph. Bernd yearned to break 250 mph, an achievement that would make him the first ever to do so on an ordinary road.

The glamour attached to these attempts—and the international attention they attracted—prompted Adolf Hühnlein to put on the show-stopper that he called Record Week. The annual event aimed to prove the Third Reich built the best roads and the fastest automobiles in the world. The general public flocked to the event, along with a host of press and party high officials.

Rudi was first to set off to break the flying-start records for the kilometer and mile. The weather was cold for October, the pavement frosted white. Newsreel cameramen and photographers caught every moment of his departure. The twenty-five-foot-wide autobahn ran two lanes in each direction between Frankfurt and Heidelberg. NSKK cars patrolled the opposite side of the highway to prevent any onlookers from moving onto the road.

As Rudi gathered speed down the band of concrete, the Mercedes felt like it was rising up at the front. At 245 mph, Rudi realized that wind pressure under the high-gloss shell had lifted the front wheels almost clear from the pavement. He could not see the road ahead over the nose of the Mercedes, and he had lost any steering at the front.

The car drifted away from its straight line. It took every bit of his ability and nerve to slow it down while keeping it pointed forward. Otherwise, Rudi would surely have flashed off the road, most likely in a terrible series of flips and pirouettes.

The attempt abandoned, he returned to the Mercedes pit by the Zeppelin hangar and informed Uhlenhaut of the issue. His team tried unsuccessfully to fix the problem, but their hopes for Record Week were dashed.

Bernd Rosemeyer was next out in his aerodynamically shelled, sixteen-cylinder P-Wagen. Since first joining Auto Union, he had been pestering his bosses to pursue world speed records. He now held several, but neither these nor his marriage to Elly, nor his impending fatherhood, had dampened his ardor for speed. His shellacking by Rudi throughout the 1937 season only heightened this desire.

As Bernd sped down the road, he held the steering wheel with a featherlight touch—adjustments beyond a millimeter or two at such velocities would be catastrophic. Every seam in the road jolted the car and left its chassis vibrating like a tuning fork. Almost every mile, there was a bridge underpass, and when he entered Bernd felt a punch to the chest from the displaced air. Then, when he exited a split second later, he needed to counteract the car's violent side swerve. Maintaining his line on the road at such speeds, often through long bends, was an experience that one writer likened to "crossing the Niagara Falls on a tightrope." At one point Bernd's heart was beating so fast that he was dizzy to the verge of losing consciousness. By the end, he was left numb and exhausted.

Over the next three days, Bernd broke a trio of world records and a dozen world class records, including long-distance ones such as the 10 Mile Record. He also clocked over 250 mph, achieving his grand ambition. It was a tour de force.

After his final run on Wednesday, Bernd called his very pregnant wife in Berlin to inform her of his string of records — or "presents." Three weeks later, Elly presented him with her own gift: a son, Bernd Jr.

The resounding success of the 1937 Reich Record Week was trumpeted by the propaganda ministry at every opportunity. "We proved once again," one editorial stated, "how great is the performance of our racers, our engineers, and our workers who build these cars." Another headline boasted, "Four Years of International Racing. Four Years of German Wins."

Neither Rudi nor Mercedes could brook letting their nemesis's records stand unanswered for a whole year, and so Kissel and Werlin began pushing Hühnlein to allow another record attempt before the next Berlin Motor Show.

Their impatience would ultimately doom one of Germany's greatest drivers of all time and fulfill the warning Bernd had once given Rudi after a particularly heated duel on the track: "We cannot go on this way . . . One of us will die."

Three days after Bernd swept up his bunch of speed records, René Dreyfus was pedaling furiously around the Bois de Boulogne, along the line of oak trees that bordered the path. Cycling beside him through the park were Louis Chiron and a couple of other French drivers. They often trained there, to keep in shape and to hone their competitive instincts chasing each other around the Longchamp Racecourse, normally used for horse racing. Some mornings, they challenged Wimille and his Bugatti crew, who also trained in the park.

During a break to catch their breath, René and his companions spoke in awe of the achievement by Rosemeyer.

A few weeks later, British driver George Eyston piloted *Thunderbolt,* a six-wheeled monster powered by two supercharged Rolls-Royce airplane engines, to a new overall speed record of 312 mph in the Bonneville Salt Flats. René and his fellow cyclists were impressed again but considered Rosemeyer's achievement superior in almost every respect. Although the silvery streamlined P-Wagen looked like it had taken its design from a science fiction novel, when all was said and done it was just a four-wheeled automobile on a concrete road. René knew that he would soon have to pit his Delahaye 145 against cars of similar engineering from Auto Union and Mercedes.

This fact struck home on November 29, when he and Jean François went to meet the Schells off a train. Inside the bustling heart of Gare Saint-Lazare, they welcomed Lucy and her husband, newly returned from their transatlantic journey to New York. They had been away since the Million and had made the trip in part to visit the Vanderbilt Cup organizers.

"Business done," Lucy said. "Écurie Bleue is set. We're going to the Roosevelt Raceway."

René was overjoyed at the news. He had long wanted to compete in America, but his departure from Alfa Romeo had extinguished that hope. Until now.

With Lucy back, focus on the 1938 season intensified every day. Public interest in the revised formula was high. According to *Motorsport,* "We shall soon be back in the palmy days when there is a whole host of new Grand Prix cars and when rumors and facts concerning their construction are freely bandied about long before race day."

On the French side, it looked like Louis Chiron might represent Talbot again. Tony Lago was reportedly building a blown, three-liter V16 engine for the new formula. Wimille would surely be back with Bugatti. The Molsheim firm had spent much of the fall attempting to beat the time René had posted in the Million (and failing). For 1938, they remained cagey as to whether they would field the car that had faltered at Montlhéry or bring out another design altogether.

The Italian firms Alfa Romeo and Maserati, long-suffering in defeat, were intent on returning to the victory stands as well. In March 1937, Alfa Romeo had bought out Enzo Ferrari's operation, although they allowed him to run the team largely as he saw fit. Even with Nuvolari on his team, Ferrari rarely posted any wins outside Italy.

For the upcoming season, Alfa Romeo hired a new overseer of its racing operations, Wifredo Ricart, who ousted Vittorio Jano from the design department and reduced Ferrari to little more than a figurehead, albeit a testy one.

Nuvolari was put under political pressure to remain on the team, now renamed Alfa Corse, despite recruitment efforts from Auto Union. "The Flying Mantuan" was losing patience with such demands, especially since the new Alfa formula car, Tipo 308, was largely a rehash of a former design. One critic remarked, "I fancy I can see some of us looking up certain old drawings and pulling some old engines off the shelf."

As for Maserati, the brothers had sold out their manufacturing operations to concentrate solely on racing and were bringing out a doubled-up version of their 1.5-liter *voiturette* engine mounted on a sleek new chassis. Achille Varzi, although hobbled by a morphine addiction, looked like he would race for them.

Even Great Britain seemed likely to be represented. The manufacturer English Racing Automobiles had announced a supercharged two-liter dynamo of a car. Its roster of drivers was yet to be released.

Most attention focused on the Germans. Rosemeyer was expected to lead Auto Union, whose new car was reported to be powered by a three-liter, supercharged V12 engine, now located mid-chassis rather than in the rear. According to early reports, Mercedes was bench-testing a similar type of engine mounted on the chassis that had performed so well in 1937. "Going on past form," *Autocar* stated, echoing the predictions of most, "the German cars are likely to prove superior again."

Shortly before Christmas 1937, Monsieur Charles departed Paris for Reims to meet with AIACR delegates from the manufacturers who intended to compete in the next Grand Prix season. Neubauer, Feuereissen, Costantini, Bugatti, Lago, Maserati—they were all there. Over a lavish dinner in the vaulted cellars of champagne producer Louis Roederer, they finalized the 1938 race schedule.

The season's start would be a round-the-houses circuit—but not

Monaco, the traditional opener. Instead, the organizers selected Pau, making the otherwise secondary race in the provincial French town one of great import for the Grand Prix. In Pau, competitors in the new formula would face off against each other for the first time. Teams would want to demonstrate their cars had what it took to dominate throughout the year. Drivers would aim to show how sharp their skills were — and how intent they were on victory. The April 10 race would truly be a proving ground.

In Frankfurt, on January 27, 1938, high winds and steady rain postponed all planned record attempts. The winter date, rammed through by Mercedes to precede the Berlin Motor Show, was always going to be a challenge. Rudi spent the day at the city's posh Park Hotel, alternating his time between Baby and Neubauer. Several Nazi officials also demanded his attention.

The bad weather delayed the arrival of Rosemeyer, who was flying his own plane down from Berlin. Elly was in Czechoslovakia, giving a lecture, and she had brought their baby son with her. It was well after dark when Bernd landed, after sighting a military airport's lights through a break in the thick clouds.

Feuereissen picked him up. They drove along the stretch of the autobahn where Bernd would defend his records if he had to. After they had dinner together at the Park Hotel, Bernd retired early to his room. Before bed, he called Elly and told her that she might well get to see him drive if the attempt was postponed until after her return to Germany. "I'm convinced that our new car will be damn fast," he added.

But the winds settled and the clouds cleared over Frankfurt just the next day, before dawn. Awake early, Rudi drove out of the city in a Mercedes coupé with Baby. A crescent moon hung in the sky. It was frigid, and everything was covered in a sheet of hoarfrost: the pine woods, the highway, the strip of grass between the opposing carriageways.

By the airport, a dance of lights broke the dark. The mechanics from Mercedes and Auto Union had arrived long before anyone else to prepare for the attempts. The focus of the day was the Flying Kilometer and Mile Records currently held by Bernd.

Rudi located Neubauer, and the two stood together in front of their Silver Arrow adorned with a swastika behind the headrest. Such was the chill in the air that their every word, whether about the car or the suitability of the weather, came with a puff of vapor cloud. In the gleam of the pale moonlight, the car looked like it might be a space-ship descended from the skies, an impression strengthened by the cock-pit domed in Plexiglas and the bubble of sheathed aluminum that surrounded each tire.

Uhlenhaut and his engineers had refined their record-breaker to provide more stability at high speeds. They lengthened the chassis and shaped the body to eliminate the lift that had nearly killed Rudi the previous October.

At 5:00 a.m., Rudi and his backup driver Manfred von Brauchitsch drove down the autobahn in a touring car. They moved at a snail's pace, looking for any imperfection in the pavement or break in the pine trees that bordered the highway where a stray gust of wind might blow through. After the one-kilometer finish, a 300-foot section of road unprotected by the screen of trees was especially exposed to any dangerous side winds.

Fortunately, the morning was almost still. Rudi was only troubled by the frost. On returning to the start, he told Neubauer that he insisted on waiting until the sun had melted away the slick white coat from the highway. Over the next hour, he paced the parking lot by the Zeppelin airship hangar where the two German teams had established their temporary pits. The sky brightened, and the sun crested the mountains to the east. The mist in the air evaporated, leaving the branches of the neighboring trees looking like stretched-out fingers. A flock of crows cawed from their heights.

At 8:00 a.m., the frost had disappeared from the road. Dressed in his racing overalls, Rudi kissed Baby for luck, then headed to the highway where the Mercedes crew had placed his car. He stuffed wax in his ears and climbed inside the cockpit. When the mechanics secured the Plexiglas dome over his head, he looked as though, if anything went wrong, he was already sealed in his well-polished silver coffin.

At 8:20 a.m., the timekeepers ready, Neubauer shouted "Go!" Several mechanics pushed the Mercedes from behind, then Rudi started the engine. He took off down the autobahn, shifting gears and gathering speed before he reached the timed section.

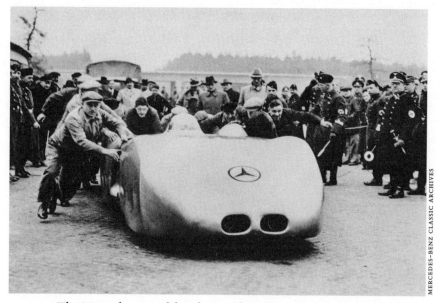

The Mercedes record-breaker needs to be pushed at the start

100 mph.
150 mph.
200 mph.
250 mph.

The treble note of the compressor pierced the winter morning. The road ahead narrowed into the slender line of an arrow shaft. The trees on either side of the road fused into a single bordering wall of black.

Rudi crossed the beginning of the timed section. He was already pushing the Mercedes to its maximum. Onward he continued, zooming down the road, covering 120 yards every second. He shot through a bridge underpass, feeling the same thump to his chest that Bernd had experienced on his attempt. Moments later, he passed the waving flag at the finish of the mile distance.

He dared not apply the brakes. At this pace, that would only bring calamity. Instead, as though letting air out of a balloon, he gradually released the pressure on his accelerator and allowed the car to roll to a stop. Several mechanics chased him down. They unbolted the dome, and Rudi gratefully sucked in fresh air. They shouted and shook his

Rudi torpedoes down the highway in his world-record speed run
for Mercedes, January 1938

hand, their voices muted and tinny as the din of the engine still echoed
in his skull.

After they turned the car around and led him back to the finish line,
Rudi waited for the telephone call reporting his time. He smoked a cigarette, his fingers trembling as the adrenaline drained away.

"It's a record, Mr. Caracciola!" someone announced.

He had completed the mile in 13.42 seconds at a colossal average of
268.3 mph.

The mechanics surrounding the car erupted in jubilation, but Rudi
remained sober. He still had to make the run back to the start line. The
official record would only be his if the mean average of the two runs was
faster than the speed Bernd had posted in October.

Minutes later, Rudi shot forward once again, accelerating until the
road became a white ribbon ahead. The car streaked so quickly that it
was moving faster than his mind could process his immediate surround-
ings. He trained his gaze far down the road to watch out for any decisions he would have to make. Aiming the car through the underpasses
felt like threading the eye of a moving needle.

Another wave of a flag. Another rush of people surrounded him. He
had bettered his time by four-hundredths of a second. His average speed
over the two runs was 268.5 mph. It was a new world class record.

"You want to do one more?" Neubauer asked.

Rudi shook his head, unable to muster words. He had already broken Bernd's record by almost 20 mph.

After navigating the cluster of reporters, he sat with Baby in the warmth of the Mercedes coupé, where she had stayed throughout the attempt. For a few minutes, they said nothing, while Rudi gathered himself after pushing so close to the brink of death.

By nine o'clock, they were back in the Park Hotel. During breakfast with Manfred and Neubauer, who was celebrating the victory with an order of sausages and Jamaica rum, a waiter came to their table. There was a telephone call for Neubauer.

When the Mercedes team manager returned, he informed Rudi that Auto Union was preparing an immediate challenge to the record attempt. Rosemeyer was already on his way from the hotel. They wanted to seize back the record, ostensibly before the next newspaper edition. Neubauer left for the track, but Rudi was reluctant to go and watch. Although he instigated this second round of record-breaking attempts, he knew they should not be run like a race. They were too dangerous already—more so if rushed—and the best time of day to run them was early morning, when the wind was at its calmest. On his way back to the hotel, he had seen a slight breeze in the treetops.

Manfred eventually convinced him they should go, and they drove out to the airport. The winds had only picked up, and Rudi was certain that the attempt would be called off. They arrived just as Bernd was coming back from a warm-up run. The Auto Union mechanics hovered around his car, making some last-minute checks. Since October, they had increased its engine capacity and tried to streamline it further.

Rudi took a look at the windsock above the airship hangar. It stretched straight out, with occasional bursts to the left. He thought it madness that Rosemeyer would try to reclaim the record in those conditions. Pushing through the throng, he approached the Auto Union driver.

"Congratulations," Bernd said.

"Thank you," Rudi returned. He thought he should say something about the wind, warn Bernd off the attempt. Auto Union could try tomorrow. Instead, he stayed silent. He would not have wanted anyone raising such doubts with him before he was about to drive. At the speeds they were pushing, nerves could kill.

Off Bernd went with a scream of his engine, and his silver car disappeared around a bend in the road. Almost half an hour later, he returned in a flash. "Good! Good!" he said to Karl Feuereissen. Over the two runs, he averaged 268 mph for the mile. His second run was much faster than the first, and he and his team thought their chances were good to retake the record on the next try.

Bernd recommended that they close off some of the air-intake ports for the radiator to keep the engine hot enough to maintain the high speeds. The mechanics made the adjustments, changed the spark plugs, and topped off the fuel tanks. Soon the car was ready.

The wind had strengthened further. To escape the cold, Rudi and Manfred sat in their car with the engine running. In any case, there was little to see during these record attempts.

At 11:46 a.m., Rosemeyer sped away down the autobahn, the din of his engine vibrating the air. Neubauer stood next to Feuereissen, listening to the staccato call from observers announcing Bernd's progress toward the starting line, which was located at the highway's 7.6-kilometer marker.

"Kilometer 5 . . . through."

"Kilometer 7.6 . . . through."

The record run had started. Seconds would determine if Bernd had done it.

"Kilometer 8.6 . . . through."

Bernd neared the gap in the trees where his P-Wagen was vulnerable to side winds.

Nobody knows for certain the cause of what happened next, whether a gust walloped the side panel of the car or whether Bernd altered his road position in anticipation of the wind. But, at over 250 mph, something went irreversibly awry.

Bernd's tires brushed the grass on the left edge of the road. The car skidded at an angle across the road, an uncontrollable slide. In an explosion of metal, the aerodynamic shell tore away. The car somersaulted several times in a blur of silver, disassembling as it went. It continued to hurtle forward, and Bernd was thrown out of the cockpit. He landed in the woods over eighty yards from the road. The car, now little more than a misshapen chassis on crooked wheels, hobbled to a stop on the embankment by the next overpass. The trail of wreckage was strewn along the highway for almost half a kilometer.

Back at the airport parking lot came the report: "Kilometer 9.2 . . . the car has crashed!" With a face like chalk, Feuereissen dropped the telephone and ran to the nearest car. The entire Auto Union team rushed after him to the site.

Bernd Rosemeyer at the start of the record attempt that cost him his life.

Sitting in his Mercedes, Rudi unrolled his window and stopped a boy who was rushing past to ask what had happened. "Rosemeyer has crashed," the youth shouted over his shoulder as he kept running. Rudi and Manfred looked at each other. They had seen enough crashes and enough death to know what awaited anyone who went out to investigate. There was no use in witnessing such scenes.

"I don't want to go there," Rudi said.

"Neither do I," Manfred replied.

Two bystanders found Bernd Rosemeyer's body resting at an awkward angle against a tree trunk. His face looked serene; his eyes were open. They were unseeing.

That afternoon, an Auto Union doctor rang Elly in Czechoslovakia to inform her of her husband's death. Soon after, news reports announced the tragedy to a shocked German public. In his diary, lamenting an already "heavy day" mired in politics and Jewish affairs, Joseph Goebbels wrote, "A deadly misfortune. Our best driver is lost in a great and completely unnecessary record race."

Nonetheless, Goebbels perceived an opportunity and staged a state funeral for his champion in Berlin. To the cadences of Beethoven's funeral march, Rudi, Manfred, and the other German drivers walked before the casket in their white racing overalls. SS officers stood at attention along the route. Hitler's own elite company of guards watched over a bereft Elly. As the coffin was lowered into the ground, the SS loyalty hymn "Wenn alle untreu werden" ("If All Become Unfaithful") was sung. Salutes and speeches followed.

The many letters of condolence to Elly from Nazi high officials were leaked to the press for maximum effect. Hitler sent a telegram, writing that "the news of your husband's tragic fate has left me shaken. May the thought that he fell fighting for Germany's reputation lessen your great grief."

Propaganda reframed every aspect of Bernd's death. He did not perish in a foolhardy test of speed; rather, as Fritz Todt, the builder of the autobahn, remarked, "He died as a soldier in the exercise of his duty." Elegiac obituaries and photo montages filled German media. A book was quickly put together. Newsreels showed both his triumphant glory on the racetrack and the ugly wreckage on the autobahn.

Bernd was likened to Siegfried, the dragon-slaying hero of German mythology. "Bernd Rosemeyer — friend, comrade . . . Your competitiveness, your chivalry, your willingness to serve, your fighting spirit. Winner of many battles. You enter the world of immortality." Rudi, whose relentless feud with Bernd had incited the January attempt, wrote his own widely distributed tribute bylined, "Your friend Rudolf Caracciola."

The tragedy did not stop Rudi from discreetly requesting his bonus — 10,000 Reichsmarks — from Mercedes for his new world class records.

There was little investigation into what caused the accident, whether it was a warped aerodynamic shell, driver error, or the impact of a cross-wind. "Fate took him from us," the Third Reich declared. No blame was assigned, at least not publicly, to Auto Union's decision to allow Bernd to make his run that late morning — or to attempt to break a speed record in the middle of winter in the first place. As Hühnlein wrote to Kissel on December 1, 1937, when initially approving the "chivalrous match," the "exceptional propaganda appeal at home and abroad" was worth the risk.

13

"Find Something"

O N JANUARY 19, 1938, nine days before the accident that killed Rosemeyer, René boarded the *Théophile Gautier* ocean liner in Marseille, accompanied by Laury Schell and their new Delahaye 135 Special. Fitted out for the Monte Carlo Rally, it boasted a high-performance 3.5-liter engine, a light-alloy body, and a full-length undershield of sturdy aluminum bolted to the chassis. Spare tires and wheels were fastened to the side of the hood, and a shovel was strapped under the rear window.

With winter storms lashing eastern Europe, their starting point in Athens looked especially brutal that year, and they needed a car to meet the challenge. Tests at Montlhéry had seen their Delahaye reaching a top speed of 112 mph, making it one of the fastest in the field.

After an extended, and much needed, rest since the Million, René was ready to drive again. Bolstered by his success, he now had the confidence to take on any competition, and the rally was exactly the kind of warm-up he wanted before the Grand Prix season began. Lucy was game, knowing how well his participation had served him in the past.

Crossing the Mediterranean was a fraught affair. The seas were rough, and on several stretches French warships escorted them through the same waters navigated by German and Italian troopships and cargo boats, often supported by submarines, bringing troops and supplies to the Fascist General Francisco Franco in the midst of the Spanish Civil

War. René distracted himself from the perilous travel by memorizing every town on their route from Greece to Monaco. After four days, he and Laury safely arrived in the port of Athens, and early on January 25, the starter flag sent them on their way.

Soon they were climbing through the snowbound mountains. It was well below zero, and their icy breath clouded the windows. Neither their fur-lined coats nor their heavy boots kept out the cold, and snow entered their vehicle through the holes for the gear and brake levers. Like sailors in a leaky boat, they were constantly bailing out snow through the windows. Still, they made it through the icy pass and to the first control point on schedule, despite their brake cables nearly freezing to the undershield.

In the mountains between Larissa and Thessaloniki, the snow only grew deeper. The Delahaye lacked sufficient clearance from the road to make it through the drifts unimpeded. Stones were mixed into many of these snow drifts, and while the two rallyers were battling through one such drift, a rock pierced their crankcase. Soon after, the needle on their oil-pressure gauge sank. A brief stop revealed an inky black trail behind their car.

In the next town, they found a mechanic's shop, but the hurried repair only stalled the inevitable. They continued to leak oil, and every few hours had to stop and refill their engine. They played this stop-fill-go game for three days before retiring from the competition.

Even if they had reached Monaco, they would have had to leave their car out overnight before the flexibility tests. Their engine would have lost all its oil, and the rules forbade adding any more at that point.

The day they abandoned the rally, they learned about Rosemeyer's fate. Interviewed soon after the news broke, Robert Benoist said, "It's a tragic accident, but that's the fate of the job." René did not care to dwell on it, knowing well the risks he was taking with Écurie Bleue.

Noted journalist Robert Daley summed up best the impact of this kind of accident on racers. "Once the horror and shock are gone he does not think about it, in the way that you and I do, at all . . . We think of it physically. We have an emotional reaction. He can think of it only intellectually: If he enters such and such a corner too fast he will probably kill himself."

There were only a few dozen drivers in the world at the time whose perspective was trained to this level to pursue victory at mortal speeds. In times past, René had lost this edge. First with Nuvolari, and now more profoundly driving for Lucy Schell, René had rediscovered it.

He and Laury continued on to Paris in their hobbled Delahaye. There wasn't much for René to do in the city. Jean François had managed to wring out only a little more speed from the 145, and René had driven it enough to know it as well as if it were an extension of his very self. Neither engine power nor driver skill was likely to prove the decisive factor at Pau. Instead, he believed that his desire to win would be all-important. That spring, in the two months before the first Grand Prix race of the season, the Nazis would give him every reason to hone that desire.

On February 18, a thin blanket of snow clung to the parks as Rudi Caracciola led a phalanx of Silver Arrows from the Reich Chancellery through the Brandenburg Gate and down a broad, tree-shaded avenue toward the Kaiserdamm, where the Berlin Motor Show was taking place. Twenty thousand NSKK troops lined their path, and behind them crowds of Berliners watched and waved. At the end of the long procession of motorcycles and cars came Hitler and his chief officials, traveling in a cavalcade of limousines that had been burnished to a shine.

Outside the exhibition hall, Rudi and the other drivers revved their engines, the yelp of the superchargers echoing against the building's marble walls and deafening the crowd. Hitler loved it, and he offered handshakes to his racing champions before stepping into the entrance hall surrounded by black-uniformed SS. Trumpets sounded as he walked toward a raised platform set in front of swastika banners that rose nearly as high as the glass rooftop. Around the platform stood dozens of NSKK standard-bearers, holding the flags of their mechanized divisions aloft.

Since early February, the Nazi leader had inflamed a world already on edge. He had forced out his two lead generals, effectively making himself the "Supreme War Lord." Then he replaced his foreign minister with Joachim von Ribbentrop. The moves signaled that Hitler was consolidating his absolute power and readying a more aggressive policy toward his neighbors, possibly even war.

In the second week of February, Hitler summoned the Austrian chancellor, Dr. Kurt Schuschnigg, to his alpine retreat in Berchtesgaden. After a tirade designed to intimidate Schuschnigg, Hitler insisted that he appoint Austrian Nazi politician Arthur Seyss-Inquart as his interior minister and provide amnesty to all Nazi prisoners in his country—or risk invasion.

When Schuschnigg balked at his demands, Hitler said, "You don't believe you can hold me up for half an hour, do you? Perhaps I'll appear some time overnight in Vienna; like a spring storm." Schuschnigg eventually capitulated, an act that a Vienna-based American diplomat believed meant "the end of Austria."

The many foreign ambassadors and dignitaries who had assembled in the Kaiserdamm for the Berlin Motor Show wondered what Hitler was planning to do next. That morning, however, he spoke only about his "beloved child," the automobile industry. After a rousing introduction from Goebbels, he came to the podium, notes in hand, and began, in a low voice. "When I had the privilege of opening the exhibition in Berlin five years ago, people questioned the value of such events." Quickly he gathered momentum, his tone rising almost to a shout, then falling to a murmur, as he charted out his resurrection of the industry and the "incomparable triumph" of its racing cars.

Fists clenched, jaw tight, he pounded the podium and looked out at his rapt audience. He spoke of the NSKK's success in training 150,000 young men to drive; he enumerated the thousands of kilometers of the autobahn that had been built; and he heralded the huge plant soon to be built to produce the "People's Car."

Although there was little applause while Hitler was speaking, there was a sense of "fellowship between the Führer and his listeners," as one French observer noted. As he neared the conclusion of his speech, fifteen minutes later, Hitler took time to lament the death of "the best and most courageous" Bernd Rosemeyer, and he announced the creation of a "Motorsports Badge" to inspire young people to be like the great Rudi Caracciola, who would be its first recipient for his "fight year after year for Germany."

Finally Hitler declared the exhibition "Open!" Behind him, curtains were swept aside to reveal the main hall. A brass band played as he stepped down from the podium to begin a tour that lasted three hours.

Set on a high square platform in the "Hall of Honor" was the stream-lined Mercedes in which Rudi had broken his world records, and indeed Rudi himself was a star of the whole show, second only to his Führer.

Two days later, Hitler delivered a speech in the Reichstag, stating that "over ten million Germans live in two of the states adjoining our border." In his view, it was "unbearable" that they did not enjoy his "protection." He was being far from coy. Seven million German-speaking people lived in Austria, and 3 million Sudeten Germans lived in Czecho-slovakia, and he had just put the world on notice that they belonged to the greater Reich.

Hitler now waited for his opportunity to unite Germany and Austria. Schuschnigg provided him with the perfect chance when he announced that he would hold a plebiscite to ask his countrymen to vote for "a free, independent, social, Christian, and united Austria—*Ja oder Nein?*" In response, Hitler issued an ultimatum to Schuschnigg: cancel the vote, resign, and appoint Seyss-Inquart as chancellor. Otherwise, German troops would march across the border.

By the time Schuschnigg submitted, Nazi mobs in Austria had already taken over the streets and town halls; German tanks and motorized columns entered the country on March 12. Hitler followed in an open-topped gray Mercedes, stopping at his childhood home in Linz. Church bells rang, and the crowds waved swastika flags and cheered the arrival of their conquering hero, who stood ramrod straight in the front seat, one hand on the windshield, the other raised in salute. He then journeyed to Vienna to seal the union (Anschluss) of the two countries. As one of his ministers stated, Hitler was in "a state of ecstasy."

Spasms of violence overtook the capital, most of it directed at the Jewish community. Mobs threw bricks through the windows of Jewish shops and looted their owners' wares. "Individual Jews were robbed on the open streets of their money, jewelry, and fur coats," historian Ian Kershaw wrote. "Groups of Jews, men and women, young and old, were dragged from offices, shops, or homes and forced to scrub the pavements in 'cleaning squads,' their tormentors standing over them and watched by crowds of onlookers screaming 'Work for the Jews at last,' kicking them, drenching them with cold, dirty water, and subjecting them to every conceivable form of merciless humiliation."

Nobody, whether in Europe or the United States, lifted a hand to stop

Hitler from annexing Austria or to prevent the pogrom that ensued. In the British House of Commons, Neville Chamberlain said, "The hard fact is that nothing could have arrested what has actually happened — unless this country and other countries had been prepared to use force." The journalist William Shirer, always in the right place at the right time, wrote plainly in his diary: "Britain and France have retreated one step more before the rising Nazi power."

Upon returning to Berlin, Hitler announced new Reichstag elections and a plebiscite of his own to be held in Germany and Austria to sanction the union that he had brought about by force. He scheduled the vote for April 10 — the same day as the opening of the Grand Prix season at Pau.

In a park somewhere outside Milan, songbirds trilled in the trees and a rabbit bounced across the grass. The rising sun evaporated away the last vestiges of dew. It was the same scene that had endured there for centuries — apart from the winding track of asphalt that cut through the park like a scar.

Then, in the distance, a shrill note rose over the landscape. The birds grew silent. The rabbit hunkered down. The sound grew quickly, penetrating the air with a raspy growl that swelled with intensity as it came closer. All of a sudden, a silver Mercedes, its hood removed to allow easy access to its engine, swept over a rise into view. Rudi Caracciola was at the wheel.

He pointed the ground-hugging arrow through a chicane and then vanished almost as quickly as he had come. Only the wail of his three-liter supercharged engine — which a reporter likened to the cry of "ten thousand scalded cats" — remained, before that too drifted away in moments.

Since early March, Rudi had left his Milan hotel every morning and driven out the highway toward Brescia. At the entrance to the park, formerly a royal estate, the gatekeeper, Biancho, would wave him past. Down a bumpy dirt road, past some stables, then a horse track, and then Rudi was at Monza, site of the Italian Grand Prix and a winter training ground. In the sunken pits, he would meet up with Neubauer and Uhlenhaut — both usually there before him — along with his teammates Manfred, Hermann Lang, and Richard Seaman. Then they would begin their day, following a rigorous preset schedule as they put the new W154 through its paces.

The V12 engine made of carbon steel was first finished in early January. Testbed readings had it pumping out 427 horsepower at 8,000 rpm. It was installed in its chassis the following month; then, in an atmospherically controlled workshop at Untertürkheim, the car was placed on a system of rollers and paced through hundreds of miles. Afterwards, they took it out on the autobahn for several long runs. But nothing would give them better information about the state of the car—and what needed refining—than the trials at Monza. The Nürburgring would have been an equally effective place to test the car, but in the winter months it was usually snowbound.

Every day they drove the car hundreds of kilometers on the Monza track and on its adjoining road circuit, often to breaking point. They tested the car's tire wear and its braking ability, acceleration, and maximum speed. They analyzed how it handled the road, the amount of fuel it consumed, how it sprang through corners, its carburation. Everything was noted in daily reports: air temperature, wind readings, distances traveled, chronometer readings on how fast a turn was taken, when a tire or brake smoked from wear and tear.

The Monza circuit provided a range of hairpins, curves, straights, and changes in elevation, but when it failed to provide the exact conditions they wanted, they created their own mini-courses, putting up pylon barriers to create chicanes, tighten corners, and shorten straights.

Neubauer and Uhlenhaut ran the operation like scientists, isolating variables such as tire width and tread, fuel blend, shock-absorber settings, gear ratios, and weight distribution in a series of controlled experiments to see what worked best; then they made alterations to the car based on that information. It was an expensive, time-consuming operation; supervised by numerous mechanics, it utilized several W154s and burned through fuel, tires, and the patience of the Mercedes drivers.

All of these experiments were carried out under strict secrecy. Neubauer set up an "Espionage Department" to secure the area where Uhlenhaut and his team worked. A guard was posted outside the entrance, and tarpaulins kept away any prying eyes. No advantage could be given away before the opening of the Grand Prix season.

Auto Union and Alfa Corse were testing their own new formula cars at Monza, and a steady number of journalists poked around the track. They saw the W154 only when it emerged from the pits.

Even so, the mood, at least inside the pits, had a familial air. Neubauer and his wife set up a canteen where the drivers and crews could get coffee or a meal. One afternoon they hunted rabbits in the park and barbecued them over an open pit. Lang, who had also cooked for the team back when he was a mechanic, was particularly adept with chicken and pasta.

In these moments, the drivers had a chance to talk about more than their impressions of the W154 or the season ahead. There was a lot going on in the world to discuss as well. Manfred was particularly pleased with the Führer's recent maneuvers, particularly because of the elevation of his uncle Walter to commander-in-chief of the Wehrmacht. Neubauer thought that having his nephew on the Mercedes team would bring them even more favor from Berlin. As for the Anschluss, the drivers believed that Hitler had made a "clever move" that promised to make Germany stronger and more respected in the world.

Rudi salutes the Führer at the German Grand Prix

The members of the Mercedes race team were not mere bystanders to these actions of the Third Reich. Whether they felt like they had a choice or not, they openly campaigned for the regime's policies. On March 26, Rudi penned (or at least signed his name to) an editorial in the Nazi Party daily, *Völkischer Beobachter,* supporting the Anschluss and endorsing the Führer for the upcoming referendum. "We racing drivers are fighters for the world-class German automotive industry. Our victories are at the same time triumphs of German engineering and workmanship. The Führer has once again given our factories the opportunity to build racing cars . . . Their unique successes over the past four years represent a glorious symbol of the efforts of our leader. For this, let us thank him on April 10th with a heartfelt Yes!"

The week this editorial was published, Uhlenhaut and Neubauer set up a course for their final trials at Monza that simulated the Pau Grand Prix circuit. They practiced on the mocked-up course over and over again. The W154 was running very well. "Outstanding!" Manfred declared at the end. "The goods," said Seaman.

Foreign journalists reporting on what they could see from the stands agreed. One remarked that the W154 engine "goes like the bomb." Another said that the Mercedes new formula cars made a "big impression . . . long, low, lean things, very fast, fine acceleration, fine road holding, and they handle well."

It was one week before Pau. Rather than have René prepare at Montlhéry—or let him recharge in the sun—Lucy Schell sent him off to represent Écurie Bleue in the Mille Miglia, pairing him with Maurice Varet, a mechanic and budding driver. To qualify the Delahaye 145 for the sports-car category, François returned the praying mantis mudguards and bug-eyed headlights.

René was keen to go: it would give him less time to brood about the opening race of the Grand Prix season. A thousand miles on a figure-of-eight course through Italy was sure to distract his thoughts. From Brescia to Bologna, up through a foggy mountain pass, then down into the Tuscan fields around Florence, René drove brilliantly. By the time they reached Rome, almost halfway through the 1,000 miles, the 145 was running in second place to a new lightweight Alfa Romeo Spider.

Then, on a twisty road approaching Bologna, a rock punctured the radiator of the Delahaye, and plumes of steam rose from its hood. Prudence would have had them quit. Instead, René and his copilot stopped every half hour to refill the radiator with water and drove like mad in between. They finished fourth—impressive given the circumstances.

René returned to Paris and prepared to leave for Pau. Lucy would not be joining the team because she was competing in a Concours d'Élegance in Cannes at the same time. She had done everything she could to see that René had his best chance to come out on top. In money alone, she had spent over 2 million francs to support the design of the 145, win the Million, and prepare for the 1938 season. Now all she could do was offer a few parting words of encouragement.

René and Lucy together at Montlhéry

One can imagine that Lucy invited René over to La Rairie to convey to him the importance of winning—and not only as a matter of their own pride. A single victory over the Silver Arrows at Pau might not change the tides of nations, but it could spark hope in a world darkening at every turn. As Jewish American heavyweight Max Baer said before his knockout fight with Max Schmeling, "Every blow to Schmeling's eyes is a blow against Hitler."

Whatever Lucy said to him, René left for Pau understanding for the first time in his career that this race was about more than who crossed the finish line first. He had heard too much of what the Nazis were doing in Germany and had seen too much of the rabid hatred Fascists showed toward him. He despised their actions and feared what they would do next. Any blow he could strike against their theories of their own superiority was one he wanted to take. Moreover, he understood that whether or not he identified as Jewish himself, this was of no account. Many people saw him as Jewish, both those rooting for him and those against him. He would never fully outrun his name. If he won, his victory would be a symbolic triumph against the Nazi thugs.

Years before, in words that had stuck with René for the rest of his life, Meo Costantini had told him that he needed to be more aggressive. In his run for the Million, René had transcended that advice. Now the rest of Costantini's words came to mind: if he truly wanted to become a great driver, he should "find something to struggle and fight for." As he departed from Paris, René knew that he had found it.

14

The Dress Rehearsal

O N THURSDAY, APRIL 7, two Delahaye trucks, painted on their sides with the bulldog mascot and the name ÉCURIE BLEUE, advanced through southwestern France. René and his fellow team driver Gianfranco Comotti, formerly of Ferrari, followed in their own coupés. In the late afternoon, they sighted Pau up ahead. The small city stood in the distance like a serene jewel.

Without listening to the news, it would have been impossible to know that only sixty miles away, beyond the jagged edges of the Pyrenees in Spain, General Franco, supported by Italian and German planes, was mounting a major offensive, barraging the Republicans with artillery in an attempt to divide their forces. Many believed that the horrors committed in this conflict, including the devastation of Guernica made famous by Pablo Picasso's recently unveiled large-scale painting of the Spanish town's bombing, were prelude to the kind of havoc that industrialized warfare would bring across Europe.

Set on a plateau overlooking the Gave de Pau River and punctuated by a high stone tower, Pau had medieval charm and a remarkable setting. In the nineteenth century, wealthy visitors poured in from across Europe and Russia and from as far afield as the United States (including Abraham Lincoln's widow Mary), and grand mansions—and even grander hotels—were established. Spas, casinos, and theaters followed. A long promenade shaded by plane trees was built on the edge of the plateau, as well as a funicular cable car to connect the aptly named Boulevard des Pyrenees to the bustling train station beside the river below.

Pau also became known as a center for sport and early aviation. The Wright brothers moved to "The Queen of the Pyrenees" a few years after Kitty Hawk, and many of the pioneering French and American aviators learned their craft at flight schools in the area.

In 1930, Pau won the honor of holding the French Grand Prix. It was run on a long 15.8-kilometer triangular circuit outside the city. To spur more interest, its organizers, the Automobile Club Basco-Béarnais (ACBB), created a shorter, more dramatic, round-the-houses circuit through the heart of the city, much like the Monaco circuit. Over its 1.7 miles, there was a baker's dozen of sharp turns and hairpins, a precipitous ascent and descent, and a few straights, none of them long. Its first year in 1933, through streets lightly covered in snow, was a success. Every year furthered the course's renown, culminating in its selection to host the inaugural 1938 Grand Prix race.

When René arrived in Pau, the streets were alive with preparations. ACBB staff set up checkpoints to control the flow of the tens of thousands coming to watch and erected barriers around the course to keep them safe. Near the start, workers readied a line of six covered grandstands, tested PA systems, positioned the competitor leaderboard, and constructed temporary food stands. Police directed the glut of traffic.

The practice sessions would begin the next day, so the finishing touches were being made to the pits and the refueling stations, located opposite the grandstands. Timekeepers established their posts around the course, and a legion of press, including journalists from a dozen German newspapers, milled about to interview anyone they could find.

René and Comotti helped Jean François and his mechanics unload the 145s. The Écurie Bleue team had few spare hands and was a shoestring operation, especially when compared to Mercedes, who arrived that same day. In a convoy of five large trucks, the Germans came into Pau like an invading army, with Neubauer their commanding general. One of the trucks was fitted out as a mobile workshop with lathe, welding plant, and even a rig to test shock absorbers. Another had a supercharged engine just in case supplies needed to be quickly transported ahead of their arrival from Germany.

Mercedes brought three drivers for the two cars, reserve engines for both of them, and enough equipment to build a whole other car from scratch. Beyond numerous engineers and mechanics, the team also included a doctor, tire and fuel experts from Continental and Shell, and several hack journalists to provide a supply of pro-German stories to the press back home.

Shortly after his arrival, Rudi told one such reporter that Pau was simply a "full dress rehearsal" for the Grand Prix season. From the intensity of their preparations at Monza to the vast organization they had brought with them to the race, he was underplaying its significance.

On the clear, sunny morning of April 8, the first of two days of practice, René stepped out of the Hôtel de France, a grand old establishment on the Place Royale. The funicular provided him with an easy ride down from the promenade.

It was a short walk from there to the pits where his Delahaye 145 was fueled and ready. Since the Million run, the team had painted its body blue with stripes of red and white that began at the nose of the hood and ran down either side of the car. Viewed from above, they created the letter "V," symbolic of their victory in the Million Franc Race. On each door was painted the number "2," René's number in the race. Comotti's car was painted with a "4," but otherwise his 145 was indistinguishable from that of his team leader.

The Delahayes drew only a fraction of the attention of the Mercedes W154s. This was the first public appearance of the new model, and drivers Caracciola and Lang had to wade through journalists and photographers to reach their vehicles.

The field of sixteen competitors promised a very fine race. Recovering from the fallout of the death of Rosemeyer, Auto Union would not be competing, but otherwise every significant race car manufacturer in Europe was represented, either by a factory team or by an independent driver. "All the colors are coming together for the first time in years," a local newspaper remarked.

Fulfilling the ambitions of the new formula, the competitors also presented a range of designs and engine sizes. Wimille was on the roster for the Bugatti team, reportedly with a new three-liter supercharged car. Three independent drivers, including twenty-one-year-old Maurice Trintignant in his first Grand Prix race, were also piloting

Bugattis, but older models. René Le Bègue, two-time winner of the Monte Carlo Rally, and another driver had entered the same type of Talbot that Louis Chiron had driven to win the 1937 French Grand Prix. ("The Old Fox" himself was absent from the team.) With the two Delahaye 145s and an independently driven 135, that was the sum of the French-built cars.

Representing Italy in red, Alfa Corse brought three drivers, including "The Flying Mantuan." The design for their Tipo 308's bodywork looked like it had been stolen from the Mercedes 1936 model. Maserati had yet to finish its new car, but three independents brought their spry 1.5-liter *voiturettes,* which might prove fearsome on the short, twisting course through the city.

According to early predictions, the Pau Grand Prix looked to "surpass in success" any before it. The winner of the grueling 100-lap race would earn a 30,000-franc prize, but more important, the prestige of claiming the first victory of the new formula.

Before the practice sessions had even taken place, Rudi and Tazio Nuvolari ranked as the clear favorites. One newspaper surmised that at Pau the two might put to rest the controversy over who deserved the title of "The World's Greatest Driver." Despite winning the Million, René received little more than a mention in the press.

After a few spins around the course, Tazio shot off under the noon sun for the first timed trial. Those who posted the fastest laps would benefit from the best positions in the starting grid on race day. Tazio delivered on his devil-may-care reputation, whipping his Alfa Romeo around the course in one minute, forty-eight seconds and beating his own record set at the 1935 Pau Grand Prix. It was an impressive run, and the Italian champion gave a toothy grin to the photographers and cameramen in the pits.

René performed well too. Feeling confident from the start, he posted a lap of one minute, fifty seconds, just behind Tazio. Lang finished third in his W154, with one minute, fifty-one seconds, then Rudi with one minute, fifty-three seconds, followed by Alfa Corse's Emilio Villoresi, at one minute, fifty-six seconds, then Comotti in the other Delahaye at two minutes and one second. Once the field had finished the first round of laps, they began the second.

Wanting to test whether Villoresi's *monoposto* performed better than his own, Tazio switched cars. Again, he took off quickly from the start, drove around the turn by the funicular, and up the Avenue Léon Say, climbing along the stone arched viaduct that supported the promenade above. After several more turns, he reached Parc Beaumont on the plateau. A spectator noticed a trail of liquid on the road after he sped by. He thought it might be gasoline. Others saw it as well, and a call was made to issue a stop signal on the circuit. By the time one came, it was too late.

Before Tazio realized the danger, his Tipo was already wrapped in flames. He had been going at such a speed that he did not see the fire for several seconds, and he only reacted when the heat reached his legs. To wait until the car stopped would mean burning alive. There was only one other option. Moving at 25 mph, Tazio pushed himself up out of the cockpit and threw himself onto the road. Flames licked his clothes as he rolled on the pavement, unconscious from the fall. Two French students rushed over to beat the flames away. The unmanned Tipo 308, now a rolling ball of fire, careered off the course into a hedge opposite a lake. By the time emergency crews could be marshaled, the blaze had consumed the car.

Nuvolari's Alfa Romeo in flames during practice trials
at the 1938 Pau Grand Prix

An ambulance carried Tazio to the hospital, where he was treated for second-degree burns to his arms, legs, and back, as well as facial abrasions caused by his leap out of the car. During a visit to his hospital bed, Rudi jokingly asked why his former teammate had not directed "the car into the lake just to cool it off."

Tazio was scarcely in a mood for humor. The fuel tank of his Tipo 308 had ruptured in one of the turns because the chassis had too much flex in it. Already unhappy with Alfa Corse and the state of its cars, he decided to retire from the team, effective immediately. Fearing that the same problem might afflict the other Tipo 308s, the Italians backed out of the race.

The field had already thinned that day. Wimille never showed up — the new Bugatti wasn't ready — and Talbot clocked such poor runs in the practice rounds that they decided to scratch themselves from the race as well. With the withdrawal of Tazio, racing circles had Rudi an even surer bet among the ten remaining competitors.

Such was the confidence of the Mercedes team that they went sightseeing in nearby Lourdes. Later Neubauer sent a report to Kissel, remarking unfavorably on the looks of the "high-legged" Delahaye, but adding that it was not to be "underestimated."

René spent a quiet evening at the hotel. He was reassured by his times but needed to maintain his ranking the next day if he was to be well positioned at the start. On the sinuous, short course, that would be critical.

The crowds gathering in Pau enjoyed another fine, cloudless afternoon for the second day of timed trials. Rudi got a better jump at the start, but René followed close behind, maintaining contact throughout the course while also studying the best lines to take in every turn. Coming out of some corners, he noticed the rear wheels of the W154 overspinning. René sensed this would be to his advantage.

Rudi finished the second session with the fastest time, one minute, forty-eight seconds. Lang matched this, though he suffered a minor crash and experienced engine trouble. Neubauer logged every lap time in his black notebook in his almost indecipherable script, as well as every issue his drivers reported in specific turns. Navigating the pits like a nimble giant, he also oversaw every tire change, refueling, and change of spark plugs.

Uhlenhaut and his forces measured every aspect of the W154's performance to report to Neubauer. They took multiple temperature readings, knowing the high 90-degree Fahrenheit day had an effect on the engine, tire pressure, and the state of the road. They analyzed the brakes, tread wear, and fuel consumption; they even drained some of the oil to be tested later in their mobile lab. Throughout the trials, Uhlenhaut adjusted the car's gear ratios to find the right balance between top speeds and maximum acceleration out of the many corners on the short course. He had yet to solve the overspinning of the rear wheels, but not for want of trying.

Nothing could be left to chance, and Neubauer kept his hawk eyes on the competition as well. René Dreyfus posted a best lap time only two-tenths of a second behind the two W154s. Comotti clocked the fourth best, at one minute, fifty-nine seconds, and then came Lanza in his Maserati *voiturette* at two minutes.

At the end of the day, the drivers left their cars in the pits and returned to their hotels to bathe away the oil stains and petrol fumes. The mechanics piloted the cars back to their garages—in neck-jerking low gear—to be wiped down and prepared for the next day. The starting grid was set, and the sun fell over a city eager for the battle ahead.

On the evening after the practice trials, Lucy visited a fortune-teller by the seashore in Saint-Laurent-du-Var, near Nice, to see if victory would be hers at Pau. She could not tolerate having to wait to know the winner.

Earlier that year, in February, the Fonds de Course commission announced that it had collected another million francs over the course of the year for the continued investment in French race cars in 1938. However, they would not be conducting another speed contest as they had done in 1937. Instead, they would examine each manufacturer's new designs—a Bugatti *monoposto,* a Talbot V16, and the Delahaye 155 *monoposto* that François was building—to determine who had the best prospect for investment. Lucy was certain that Delahaye would receive some of the funds, particularly given that they had won the Million. There were rumors, however, that Tony Lago had schmoozed his way into the commission's good graces—and that Talbot was likely to receive the entire sum. Lucy was disturbed by the scuttlebutt and took it very personally. Had she not done enough to prove her worth? Was a first at Pau needed as well?

One would have needed to swear off newspapers and café gossip completely not to know about the big Grand Prix race that weekend, and Lucy was far from an unknown figure on the Côte d'Azur. Perhaps the fortune-teller knew her by sight, or it may indeed have been her powers of divination. Whatever the source of the fortune-teller's knowledge, she was uncannily close to the truth.

The woman asked Lucy if she owned a stable of "racing things." Lucy nodded, lost for words. "Horses?" The fortune-teller continued. "Mechanical horses?" The question might well have knocked Lucy from her chair. She nodded again.

One of them was going to win the next day.

Lucy hurried to find the nearest phone to call Laury in Pau. She was utterly convinced that they would beat the Silver Arrows.

In Stuttgart, Wilhelm Kissel waded through the reams of reports that were being relayed by telephone from his racing team in Pau. Nuvolari was definitely out. "This takes away our most dangerous opponent," read one dispatch. "But the Grand Prix is far from being decided in our favor. In today's training, Dreyfus for Delahaye was able to stay close to Caracciola."

With the loss of so many competitors, Pau looked to be a "a fight only between Mercedes-Benz and Delahaye." There were other concerns from Neubauer and the race team as well. Dreyfus probably would not need to refuel, while the Mercedes drivers would. Their three-liter supercharged engine gobbled up the noxious WW fuel at a rate of 1.6 miles per gallon. Also, the high temperatures at Pau made the asphalt spongy and slick, causing excessive tire spin. Finally, Uhlenhaut needed to replace the engine on Lang's car. The oil pump was faulty, and the spark plugs continued to get mucked up during practice. They would work overnight to make the switch. Despite these issues, the reports stated, Caracciola and Lang were pleased with the W154s. "Flawless" was the term they used. There was every reason to believe they would win.

In this opinion, the drivers were well supported. Over the past week, the Daimler-Benz public relations department had pumped out press releases highlighting the "mastery of German engineers" and the supremacy of Caracciola. Scores of newspapers parroted the same and forecast a win by the Silver Arrows in the opening of the new formula. After all, neither a French driver nor a French car had managed to beat the Germans in any Grand Prix event since Hitler first funded their racing teams.

Looking further back in time, a historian noted that the French had not "defeated the very latest thing in Mercedes racing cars [since] the Sarthe Cup at Le Mans in 1913, and before that one had to go back to the early Grand Prix and Gordon Bennett days." Even the patriotic French publication *L'Auto* begrudgingly predicted, "Surely, Mercedes should win."

One newspaper best summed up the forecasts for Pau: "The outcome of this race is now such a complete certainty that no bookmaker would take any money on the Germans, even at heavy odds-on prices . . . The Germans don't make the kind of mistakes peculiar to the average mortal, and with Caracciola and Lang driving, what is there to add?"

Kissel was depending on a Pau win, especially since it was being held on the same day as the plebiscite in Germany to vote on a single, but conflated, question: Do you approve of the Nazi list of parliamentary candidates and the greater Reich's first conquest? Hitler had barnstormed Germany and Austria to rally support for what was already a foregone conclusion. Goebbels made sure the ballot was printed with a huge circle to mark "Ja" and a punch-hole-sized one for "Nein."

In support of the referendum, Kissel had done his own proselytizing over the past two weeks. He tied the success of Mercedes and its racing team to his "beloved leader" and reasoned that Austria, "our brother," should return to the family to benefit from the same. "Hitler led the German automobile industry away from the edge of the abyss and paved the way for its unprecedented rise," Kissel said in one of his speeches. "For this we must thank him — whether worker, engineer, salesman, technician, or director — on April 10, 1938, with all our hearts through our 'Yes!'"

The company also publicized a testimonial from an Untertürkheim mechanic, designed to show that the common man was also behind Hitler. "How different has our racing success been since our Führer rose to power," he said. "We owe it to him alone, and that is why there is only one answer for every worker on April 10. Yes!"

In a final speech, Kissel remarked that "victory in the Grand Prix was not only a victory of the driver and builder in question, but always at the same time one of the whole nation." Defeat at Pau would be a humiliation.

In the Hôtel Noutary, a diminutive establishment in Pau known for its fine restaurant, René sat down for some dinner and a bottle of the local Jurançon wine with his team: Monsieur Charles, François, Comotti, Laury Schell, and the handful of Écurie Bleue mechanics. A few reporters also pulled up chairs to join the banter at their table.

They discussed the meticulously organized Mercedes operation and the speed of their new formula cars. "Awesome" was the word most often used.

"So you think you're going to beat the Germans?" a reporter asked René.

"Well, you never know," he demurred.

A waiter approached the table to inform Laury that he had a telephone call. His wife. He stepped away to take it in the hotel lobby, visible from their table. Whatever Lucy was saying was making him laugh. On returning to the table, he told René that Lucy would like to speak with him.

Phone in hand, René told her that he was sure the next day would go well.

"I know you will do well too," Lucy said, her voice sure.

When he rejoined the table, everyone was whispering conspiratorially and trying to stifle their smirks. René tried to work out what was funny, but they refused to say. He did not press it.

Later that night at the Hôtel de France, he paced his room, his mind churning over the upcoming race. Since first becoming a Grand Prix driver, he had not beaten Rudi Caracciola in a single race. Compared to the Delahaye, the German's W154 was faster off the line, produced almost double the horsepower, and benefited from far more advanced brakes and suspension. From Neubauer down to the mechanics, Mercedes boasted a professional operation that was exhaustive in its preparations and lightning quick in the pits.

Nonetheless, René had spotted three weaknesses of the Mercedes cars during the timed trials that would play well in his favor. First, they would need to refuel midrace while he would not. The Delahaye had a large gas tank and burned through fuel much more economically.

Second, the W154s often suffered from wheel spin in the corners because their engines were so powerful. Too much throttle, and Rudi and Hermann Lang would lose traction—and time—throughout the course. In contrast, the Delahaye hugged the pavement coming out of turns.

Third, René had never shifted out of third gear on the longest straight on the course. It was simply too short to gather much speed. One hundred miles an hour was probably the maximum one could take it at. Again, this meant that the W154s would not be able to use speed to their full advantage, as they could on faster courses like Mellaha and AVUS.

René boiled down these observations into a clear, simple strategy: stay tight with the Silver Arrows; never allow them to gain too much distance; and finally, when they need to refuel, seize the lead and never give it back.

In the Million, René had proved to himself that he could run an aggressively disciplined attack and that his Delahaye was equal to the task. Pau was a real race, however, one where he would have to parry with other competitors through the city on a narrow course. Over one hundred laps, a single mistake entering a turn, a single mechanical failure, or a single stroke of bad luck could ruin his chances at victory—or worse. Unlike with the Million, there was no returning to the start to try again. Any setbacks would have to be addressed by pushing harder and faster.

Not since his first Grand Prix victory at Monaco had René wanted to win more. In 1930, he had chased the checkered flag solely for himself. The world had changed, as had he, and now there was so much more to fight for.

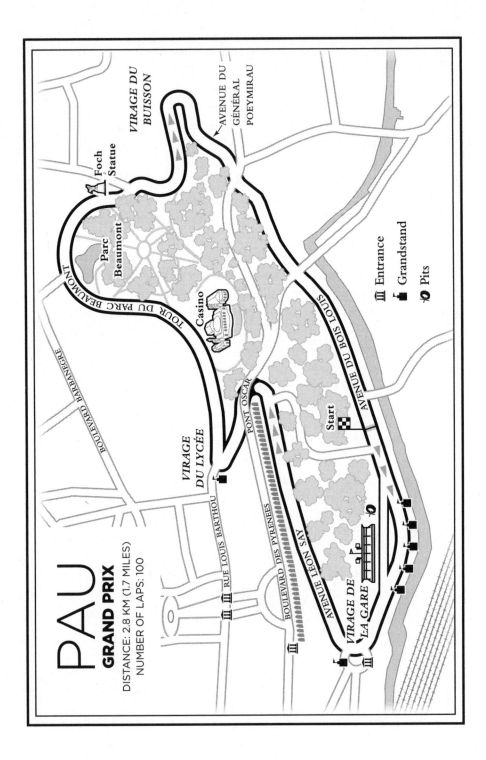

PAU
GRAND PRIX

DISTANCE: 2.8 KM (1.7 MILES)
NUMBER OF LAPS: 100

VIRAGE DU BUISSON

AVENUE DU GÉNÉRAL POEYMIRAU

Foch Statue

Parc Beaumont

TOUR DU PARC BEAUMONT

BOULEVARD BARBANEGRE

Casino

AVENUE DU BOIS LOUIS

VIRAGE DU LYCÉE

PONT OSCAR

Start

RUE LOUIS BARTHOU

BOULEVARD DES PYRENEES

AVENUE LÉON SAY

VIRAGE DE LA GARE

Entrance
Grandstand
Pits

15

Victory at Pau

T HE DAWN SUN cast the snowcapped craggy peaks of the Pyrenees in hues of gold and pink. As on countless mornings past, early risers walked their dogs along the promenade or watched the light change over the horizon. Across the cobblestone streets, the scent of freshly baked baguettes wafted from boulangeries. Espresso machines gurgled and hissed in cafés. A bell tolled. A cart laden with vegetables tottered down an alley beside a restaurant. The caretaker of the Château de Pau swept the steps with a grass broom, the *swish-swish-swish* as lulling as the tick of a clock.

Slowly, but with a gathering momentum, the city awakened. A regular flow of cars and pedestrians converged on its center. One spectator came dressed as Adolf Hitler—oily comb-over, toothbrush mustache, and all. In the sidewalk cafés and hotel lobbies, the gossip was over the latest predictions on the race. Would Nuvolari compete after all? Did you see those Silver Arrows in practice, and hear their engines? Might the Million Franc Delahaye have a shot?

Some, even if they dared not admit it, were thrilled at the thought of witnessing an accident. As Rodney Walkerley, who wrote for *Motor* under the nom de plume "Grand Vitesse," remarked, "It was as if the multitude exhaled a nervous tension which became palpable to those who realized that the actors in the great drama of skill and speed might at any moment become figures in a tragedy. For Grand Prix racing was, like all motor racing, balanced on the very brink of death."

That morning, while the other teams focused solely on the competition ahead, thirty-four members of the Mercedes team, including the drivers, gathered in their hotel to vote in the referendum. To the conflated question — "Are you in agreement with the reunion of Austria with the German Reich as effected on March 13, and do you vote for the list presented by our Führer Adolf Hitler?" — they unanimously responded "Ja!" A Mercedes official forwarded their tally to Untertürkheim by telephone. The team, some of whom had already labored overnight to install the reserve engine in Lang's car, then went back to work.

Spectators who wanted a particularly good viewpoint had lined up early in the morning, and by noon roughly 50,000 people packed the grandstands and ticketed enclosures about the round-the-houses circuit. It was a fraction of the number who attended the Nürburgring, but the much shorter Pau course made the crowd similarly dense. The heat had broken since the practice session, and there could not have been a more ideal day for a race.

Meanwhile, the pits hummed with activity. Crews inspected their cars, which they had driven down from their overnight garages. They stacked towers of tires, rolled out drums of fuel, rigged their force-feed lines, and set out toolboxes — all before their drivers arrived. Like surgeons before an operation, they positioned every supply and tool they might need in its proper place. Each car had its own slot, the name of the driver hanging on a sign beside the open-air paddock.

Most of the drivers wore a talisman or adhered to a particular ritual for luck. Rudi ate an underdone steak, then put on his white silk overalls, black belt, and, most important of all, the old, grease-stained leather race shoes that Baby likened to his "holy relic." His teammate Lang nailed up a horseshoe in the pits. Nuvolari was not competing, but if he had been he would have worn his famous yellow jersey and golden tortoiseshell necklace.

In his room at the Hôtel de France, René followed his own ritual. He took off his wedding ring and watch, leaving them on his bedstand to collect after the race. Their removal was an acknowledgment that he might well die that day. He triple-knotted his shoes, then clipped off the ends of the laces to prevent a snag on a pedal. He seemed focused on stripping away everything inessential. Then it was time to go.

By the grandstands, loudspeakers played music, stirring up the already charged crowds. Officials paraded back and forth by the starting line. Timekeepers double-checked their watches. Gendarmes pushed back against the throngs eager for a better sightline. Scores of journalists angled for interviews with the teams, photographers bustled around in the pits, and radio announcers readied their broadcasts. Only the national flags of the competitors, limp in the windless afternoon, lacked vitality.

Either the mechanics or the drivers themselves warmed up the cars on the course, the sharp roar of their engines an early taste of what was to come. When they came back into the pits, exhausts sputtering, the crews made some last-minute checks, then draped the hoods with heavy blankets to keep the engines warm.

One of the nine cars still slated for the race was missing—and it was one of the Silver Arrows. The Mercedes mechanics had thought that the engine they replaced in Lang's car was working properly, but then, during the final check, an hour before the 2:00 p.m. start, Uhlenhaut discovered a problem with its oil-circulation pump. It was impossible to fix it in time. Neubauer informed race director Charles Faroux about the withdrawal.

From the original sixteen, the field had been cut in half, and for those in the know, who had either watched the practice sessions or read the flurry of accounts in the newspapers, this meant that the Pau Grand Prix would probably be a two-horse race:

Dreyfus versus Caracciola.

Delahaye versus Mercedes.

France versus Germany.

Both Nuvolari and then Lang had dropped out, and none of the other drivers had approached their lap times in practice. To the ACBB organizers, this was a disaster. To many others, however, it created the possibility of an epic duel.

At last, all the drivers were on the scene. Autograph-seekers shouted for attention, their red-and-white race programs in hand. There would be time for that after the race.

In the Delahaye pit, Jean François reconfirmed with René their strategy: stick to Rudi until he needed to refuel, then take the lead and never give an inch of it back. Laury then drew René aside to tell him about the fortune-teller Lucy had seen, the source of their amusement at dinner the night before. He had been reluctant to tell René, believing it might unsettle his already fraught nerves, but had then decided it was a secret he should not keep. On hearing of the premonition of his sure win, René said wryly, "You should go and tell that to Rudi."

The Mercedes driver stood next to Baby in the pits. She was holding their pet monkey, Anatol, who was wearing a sweater she had knitted herself. The little rascal was another of Rudi's lucky charms.

There was no need to go over strategy with Neubauer. Rudi intended to shoot off like a rocket from the start and gain an unassailable gap from Dreyfus, regardless of the need to refuel at the midway point.

As 2:00 p.m. fast approached, a band of trumpeters played a traditional Palois tune. To the strains of the melody, the mechanics pushed the cars out of the pits and guided them toward the grid, located roughly 100 yards in front of the grandstands. White marks on the pavement indicated where the cars should stand. Rudi's Mercedes was on the left of the front row, René's Delahaye to its right. Staggered behind them in three rows were the rest of the competitors.

Dressed in a summer suit and light hat, Faroux called the eight drivers together in a circle to advise them on the rules, including to give way on the course if the marshal waved the blue flag. Probably only Trintignant was rookie enough to need to listen.

After the briefing, the drivers milled about the pits. Their crews tried to distract them with small talk, but it was pointless. Now that they were approaching zero hour, there was little to think about apart from the race. Some of the drivers chain-smoked; others adjusted then readjusted their gloves or wiped their goggles for the tenth time. No one stood still.

"Five minutes," Faroux shouted. "To your cars."

Rudi limped to his Mercedes, his gait uneven since his accident at Monaco five years before. Around his neck and tucked neatly into the top of his overalls was a polka-dotted scarf to protect from the seep of road dust.

Wearing sunglasses against the glaring sun, René headed to his Delahaye, polishing a new pair of goggles—he never wore the same pair twice. He fitted his linen cap over his head, slipped on his tan leather gloves, then pulled the goggles over the top of his head.

As he and Rudi climbed into their cockpits, their wives watched from the pits, stopwatches at the ready. Baby once wrote of those seconds before a race: "My eyes fill with tears. It is a fearful minute to pass, and I feel like praying to God Almighty to stop this folly by some miracle."

In the grandstands, the loudspeakers went silent and a hush fell over the crowd, now all on their feet, gazing at the cars, waiting . . . waiting . . .

On the grid, René eyed Rudi. The German driver returned the look. The other drivers shared their own glances: some grinning, others stone-faced.

Then, like a thunderclap, one car roared into life. Then two. Then all eight. The air seemed to shudder, and the clamorous beat of the engines took on a heartbeat of its own in the chests of everyone close by. The drivers tapped on their throttles, quickening the beat, as the track cleared of mechanics and beribboned officials. The air thickened with exhaust, whose "various dopes," one reporter wrote, "suggested boot polish and tinned pineapple."

Rudi rechecked his gear level, his tachometer, his foot pedals, his rearview mirrors. All fine, he rubbed the side of his game leg, knowing well the pain that would come with 100 laps of the round-the-houses circuit.

René sat upright in the cockpit, straight and tense as a board. He trained his gaze on Faroux. Standing on the front right edge of the grid, the race director turned his wrist to read his watch. It was time. He held the tricolor flag aloft in one hand and raised the other to count down the seconds.

René tightened his grip on the gearshift and revved his engine. The Delahaye throbbed around him, and the cacophony sounded like the final movement of a stirring symphony.

Five . . . four . . . three . . . two . . . one! As the crowd swayed forward, necks craned toward the start and Faroux whipped down the flag. René released his clutch, and the Delahaye leaped ahead like a cat freed from an open cage. Rudi burst away quicker, the tires of his Silver Arrow searing parallel black scars onto the pavement. He left a cloud of burned rubber and dust in his wake.

The surging jam of the other six cars followed in a pitched screech.

MERCEDES–BENZ CLASSIC ARCHIVES

The start of the 1938 Pau Grand Prix

René shifted quickly, his tachometer needle spiking then falling in rapid succession as he sped down the tail end of the Avenue du Bois Louis and past the first sections of the grandstands.

Into the slight right bend by the pits, Rudi was two car lengths ahead. He angled across the road to close off René from passing him. There was another brief patch of straight, then the sharp turn at Virage de la Gare, where René clung to the right curb.

As he came out of the turn, the tail of the Delahaye waggled, but he recovered and punched his accelerator, starting the climb of the long but narrow Avenue Léon Say. Such was the rush of air that his head was forced back against his seat.

Rudi streaked ahead of him, his superior power devouring the uphill straight. To their left, clinging to the side of a wedge-shaped hill dotted with palm trees, a mass of spectators witnessed the opening drama. Above them, bent over the edge of the promenade, another swell of people watched. In their light-colored summer clothes and hats, they were all but a blur of white to the drivers.

Rudi and René separated themselves from the others very quickly on the ascent. They crossed underneath a black wrought-iron pedestrian bridge that spanned the avenue. Those lucky enough to have a spot on the bridge watched the race cars pass under their feet with a shuddering boom. The avenue led toward the stone arches that supported the Boulevard des Pyrenees. Rudi was first into the left-hand hairpin that passed underneath the Promenade Pont Oscar. Such was his speed as he entered the turn that his wheels skipped across the pavement.

René followed him into what was effectively a short tunnel, his eyes allowing only an instant to adjust to the dark before he was again in piercing daylight. Already, his initial nerves had settled. The road continued to climb, and in another brief bend he ripped through a rapid change of gears. The yawp of his engine ricocheted off the street's stone buildings.

Just as he was gathering speed, he prepared to brake in front of the imposing gray edifice of the Lycée Louis-Barthou. The Virage du Lycée was so sharp a hairpin that drivers had to take it almost at a crawl. As Rudi came out of it, his Silver Arrow suffered what one reporter called "paralyzing wheelspin."

René and Rudi were now bunched together nose to tail. A missed shift of Rudi's gears and the Delahaye would ram straight into his rear.

The Mercedes swept swiftly around the Tour du Parc Beaumont, a U-shaped route around the park. This level part of the course was packed with spectators, held back by a wooden barrier that bulged with their numbers. Next, the two drivers passed the casino, an expansive cream-colored building with twin steeples. Then they roared around a looping bend lined with stacks of straw bales. The swans that usually populated the man-made lake at the tip of the park had wisely abandoned the area.

The second half of the park leg featured a right-left-right chicane that left the cars' tails zigzagging across the pavement. The section was watched over by an imposing statue of Marshal Ferdinand Foch, the Allied commander whose September 1918 offensive brought World War I to an end. The French general did not blink at the sight of a German driver in the lead.

Soon enough, Rudi was hurtling downhill. René tried to keep close, thinking not only about the sloping turn he was traversing but also about the next moves he would have to make on the constantly changing course. In that, a racing driver is not unlike a billiards player, taking a shot while simultaneously setting up the next one . . . and the next.

René followed the Avenue du Général Poeymirau along a line of houses perched on the edge of the descent until the Virage du Buisson.

At certain moments, it looked like he and Rudi were driving in opposite directions as they took the hairpin bends down the steep hill. Then, before they could gain much speed, they downshifted and jogged through another chicane.

The road never seemed to straighten, and it was so narrow that René knew that even a small error in movement would see his tires clip against the curb and his Delahaye swung into a stone wall. He escaped the chicane unscathed and entered the longest straight on the course: Avenue du Bois Louis. By the time he hit 100 mph, zipping down the tree-lined road toward the start, Rudi had distanced himself by almost twenty yards. This represented a fraction of time, but to those in the mostly French-populated grandstands, the separation was a dispiriting one.

Rudi finished the first lap in one minute, fifty-two seconds; René was a second behind. Comotti in his Delahaye, never the equal to the Million car his team captain drove, was third, nine seconds back. The others trailed in the distance.

Ninety-nine laps to go.

Again Rudi rounded the circuit in front. Again René finished close behind. They were so close at times that their cars almost touched. As René had experienced in the past, the exhaust from the secret brew of German fuel belching out of the Silver Arrow made him feel like he was being slowly anesthetized, particularly since the circuit through the streets of Pau trapped the fumes. Still, he wanted to remain tight on Rudi in case Uhlenhaut had come up with some surprise change since practice to counteract the W154's poor handling of corners.

By the end of the fifth lap, René had his answer: there had been no change. Rudi continued to have trouble on the corners. His engine was simply too powerful for the course. Given that their average speed was only 55.8 mph, he had brought a cannon to a close-quarter knife fight.

In the seventh lap, René decided to try to overtake, to see if it could be done. Accelerating out of the Virage du Lycée, he gained ground on the Silver Arrow. Around the park, he drew up on Rudi's tail, waiting for his moment. Then, as they took the wide bend that began the downhill run, René struck. He threaded through a gap on Rudi's inside and grabbed the lead. The nearby crowds cheered madly, and when the report came over the loudspeakers by the grandstands, they too joined the uproar.

Paris-soir

C'EST DREYFUS QUI, PILOTANT UNE VOITURE FRANÇAISE, A ENLEVÉ HIER LE GRAND PRIX DE PAU. VOICI DREYFUS, MENANT DEVANT SON RIVAL CARACCIOLA

ADATTO ARCHIVES

René vaults around a corner and into the lead over Rudi Caracciola

With open road ahead of him for the first time, René stamped on the accelerator and rounded the park. It was sweet to breathe fresh air again. His Delahaye was braking well. The engine, tuned perfectly for the course, was running steady. He began to think that he had a good chance of winning the race now. He had proved that he could hang tight to the German's silver bullet of a car. He had proved that he could overtake it. Now all that was left for him to do was to simply drive the race of his life. Anything short of that, and Rudi, whose style was consistent as a Swiss watch and who rarely made a mistake in either judgment or performance, would come out on top.

Over the next several laps, René held the lead. Each time he passed the grandstands, the crowd was on its feet, hats waving. The Silver Arrow was forever looming over his shoulder in the rearview mirror, tracking him like a predator around the course, nipping at his tail. No doubt the German star was trying to force a mistake or to push René into overtaxing his Delahaye.

They now had to manage interference from the other competitors, who they were beginning to lap regularly. The Maserati *voiturettes* driven by the Italians Antonio Negro and Dioscoride Lanza were struggling on the course. Negro had already stopped twice in the pits, and by lap ten the two leaders had already passed him three times. On the slender, twisting course, populated with blind turns, the slower cars were dangerous: constantly moving obstacles that could seemingly materialize out of nowhere.

In the twelfth lap, Rudi crept up to René's side, and the two almost locked together as they swept around the course, neck and neck, neither giving way to the other. They were like two gladiators in fixed combat before a crowded forum of spectators. Coming down the straight to finish the fifteenth lap, René kept to a moderate pace even though he wanted to shift into fourth gear: Too much speed tempted fate. Rudi pounced on the opportunity and overtook René's Delahaye, then almost immediately lapped Comotti in his.

A groan of disappointment coursed through the stands, while the Mercedes pits crowed. Nothing and no one could rival their dominance in the straights. Rudi looked like he was ready to leave the French upstart in his wake for good.

Once around the Virage de la Gare, Rudi barreled up the hill, the banshee shriek of his superchargers ripping through the crowd. He quickly widened his lead. There was no question: Rudi was making his move, wheelspin be damned. At the end of the sixteenth lap, he flashed down the Avenue du Bois Louis. Timekeepers punched their stopwatches and marveled at the reading: one minute, forty-seven seconds. It was a new lap record for Pau.

The Delahaye crew posted on a board so René could see that he was six seconds behind. By the faintness of the Mercedes engine, he already knew that he was completely out of contact. Perhaps Rudi had been holding back after all. By the time René passed the grandstands, the Silver Arrow was already streaming up the Avenue Léon Say. If he lost so much as a second per lap from his rival, the race would be over, even with the need for the Mercedes to refuel. If he lost two seconds per circuit, Rudi would lap him completely.

René fought to bridge the gap, but by the twentieth round Rudi maintained a five-second lead, averaging a pace of one minute, forty-nine seconds. Comotti was third, Trintignant fourth, and Raph (in a Maserati) fifth—all three a distant lap behind. Yves Matra, in his Bu-

MERCEDES–BENZ CLASSIC ARCHIVES

Rudi roars up the hill at Pau, far in the lead

gatti, had been passed twice by the leaders. Negro had dropped out, his Maserati simply not up to the task. As predicted, the Pau Grand Prix had become a contest between Mercedes and the Million Franc Delahaye, and the German car was showing better mettle.

Over the next five laps, Rudi widened his advantage, gaining the average of a second's gap that René had dreaded. He was maintaining his pace while René was struggling to keep up. Often the Silver Arrow was completely out of sight, and René was unaware of how much of a separation there was until he ran past the pits where a signal board showed him the expanding, soul-crushing gap. During one stretch, he was almost 400 yards back—to the great delight of the German team.

By the thirtieth lap, almost an hour into the hundred-lap race, Rudi had gained another second over the Delahaye. His lead was commanding, but not yet insurmountable, because of his expected refueling somewhere near the midpoint. To take advantage of it, René would have to avoid mechanical trouble—a ruptured oil line, a cracked pipe, a faulty brake drum—to say nothing of the ever-present possibility of a crash.

The course demanded a torturous routine of gearshifts, braking, jerks of acceleration, rapid switches of steering, and lurching slowdowns. It was tough on the cars and tougher on the drivers. A single instant of lost focus and disaster might strike. Both drivers had a long battle ahead before either one could claim victory.

In the press box opposite the grandstands, Karl Kudorfer, a Mercedes publicist, telephoned Stuttgart. After describing why Lang had scratched out and detailing the progress of the race, he concluded, "Caracciola and Dreyfus are at a great distance atop the leadership. Nothing can yet be said about the outcome."

In Cannes, Lucy learned the same from a radio broadcast, but it was unlikely that this sapped her faith in the fortune-teller's prediction. Lucy's confidence was an unshakable force.

Throughout Europe and beyond, motorsport fans huddled around radios in their homes or local bars, wondering what might come next in this two-man struggle. Among them was a car-crazy fourteen-year-old Jewish kid named John Weitz, who had been sent away from Berlin by his parents to study in the safety of London. He was one of many rooting for the driver called Dreyfus to win.

All through René's Grand Prix career, racing aficionados had labeled him a "scientific driver," one who typically drove with intelligence, finesse, patience, and tactical acumen. He was also known as a driver who rarely forced his car to the breaking point and who always avoided outrageous risk, particularly after his near-death escape at Comminges in 1932.

In the thirty-first lap, René surprised them all by charging hell-bent after Rudi. There was no other choice. If he continued to lag behind, the race was already over. So he began pushing himself, all but calling out his every fervent move:

Blaze up the hill to Pont Oscar. Swerve into the sharp underpass.

Faster.

Brake before the sharp turn at the Lycée Louis-Barthou.

Faster.

Press the foot flat along the long spoon turn of Parc Beaumont.

Faster.

Zigzag through the chicane past the statue of Marshal Foch.

Faster.

Swoop into the curved downhill of Avenue du Général Poeymirau.

Faster.

Charge deep into the Virage du Buisson. Stick the hairpin.

Faster.

Dive into the serpentine descent toward Avenue du Bois Louis.

Faster.

Launch into the start-finish straight. Take the higher gear.

Faster.

In a ruthless sally that captivated spectators and journalists, René closed by almost half of Rudi's eleven-second lead. He nearly bested the newly set lap record while he was at it. As he raced by the grandstands, he glanced at Chou-Chou and Baby, who had taken up position on a dollop of a hill by the turn at the train station. Sitting on the grass, his wife looked thrilled. Baby, who up until then had been all smiles under her Panama hat, was dour.

Over the next nine laps, Rudi tried to shake off René again. Every time they passed the pits, Neubauer was there, standing on the edge of the circuit, frantically waving his red-and-black flag that signaled his driver to increase his speed.

The distance between the two cars seesawed back and forth, Rudi stretching his lead in straights, René clawing it back in the corners. It was like they were tethered by a spring that would not allow either of them to draw too far away from the other.

By the fortieth lap, in a virtuoso driving performance, René had chipped down the Silver Arrow lead to only three seconds. Comotti, their closest competitor, was two rounds of the circuit behind them. In last place, piloting a Maserati, was Lanza, who had been lapped thirteen times.

As Rudi entered the forty-second lap, his pit crew continued to signal him to go faster. Flushed red in the face, Neubauer waved the flag with such fervency that it looked like the stick might snap. They both knew that the Mercedes car needed at least half a minute to refuel, not including the precious seconds lost entering and leaving the pits. The race was almost at the midpoint, and their team had only a slim lead, far from what they needed.

Rudi felt like there was nothing more he could do. On exiting turns, his wheels kept spinning out, especially now that the circuit was slicked with oil and rubber from the other cars. He could never use the full potential of the car's power in the straights. His hip hurt from all the shifting and braking. His calf felt like it was being burned raw by the exhaust pipe. Third gear kept slipping out. He should be winning this race. He was expected to win this race. For the Reich. Losing was not an option.

Again Rudi tried to widen his lead.

René hounded him at every turn. There was no need to pass the Mercedes. Its pit stop would do the work for him. Through to the fiftieth lap, he allowed the Silver Arrow a couple more seconds of a lead simply to be free of its noxious fumes. Such was his control and rhythm on the course. Such was his confidence.

Rudi had tried to break him. René had broken him instead.

In lap fifty-two, the Mercedes finally sped into the pits, and the Delahaye passed into the lead. A roar erupted from the grandstands.

With a hiss of his brakes, Rudi stopped the car. The crew sprang into action to refuel the car. After unscrewing the steering wheel, he climbed out of the cockpit. Neubauer was quickly beside him for a report on the car's performance. Instead, Rudi informed him that Lang should take over. He was done. The course was too tough on his leg. His calf was seared. Third gear was impossible. He couldn't continue.

An unusual panic spread in the Mercedes pit. Lang was there but dressed in his street clothes. Although the rules allowed such switches, there had been no reason for him to be prepared to go since Rudi rarely dropped out of a race. Neubauer failed to persuade him to change his mind, and Lang dived into his racing overalls. While he scrambled into the cockpit, the Mercedes was topped out with fuel. Lang fixed his cap and goggles, then darted onto the course. Over a minute had passed.

When René came back around toward the pits, a lap board posted his sizable lead: one minute, nine seconds over . . . Lang. Rudi had withdrawn. René was certain he knew why: Rudi could not bear to be beaten by a French car, let alone one piloted by René Dreyfus and all that was implied by that surname. Rudi might not have been an anti-Semite, but he knew that the Nazis would be displeased with their champion for letting himself be beaten by a Jewish driver.

Only spurred further by the switch in drivers, René pressed harder than ever before. The race was barely half over. The year before, Lang had won Tripoli and the AVUS race, besting his more experienced teammates. He was a bold competitor, and critically, he was fresh of limb and concentration. René had been driving for one and a half hours on the exhausting course, making scores of calculations each lap on how swiftly to brake, when to shift, how to angle into each and every turn. Sweat beaded on his forehead and soaked his chest and thighs.

Up Avenue Léon Say. Through the underpass. Into the Virage du Lycée. René eked out hundredths of seconds with a slightly sharper line and quicker shift through the gears. As he rounded Parc Beaumont, the air rang with the sounds of his compatriots urging him onward. After the long bend around the park, he took the swift right-left-right jog through the chicane. A sliver of hesitation with the brakes here, a little more throttle there. Faster. Then he hurled the car downhill. He barely needed to think: his navigation of the course — and every calculated shift of his hands and feet — was habit now. He lurched out of the Virage du Boisson, tap-danced through the next chicane, then swallowed up the stretch toward the start.

In lap sixty, René posted one of his fastest circuits of the Pau course and expanded his lead on Lang to one minute, twenty seconds. At this stage of the race, he had to gain two seconds for every lap until the finish to catch up with René. A *L'Auto* reporter predicted, "There is little chance that victory will escape him."

Kudorfer had much the same message when he next telephoned Stuttgart: "It seems very doubtful that Lang will catch up with them if Delahaye holds."

The booming cheers of the spectators — and the excited staccato commentary from the French broadcasters — were far less restrained. René was absolutely brilliant; his victory looked assured. One could picture Lucy hollering and whooping in Cannes.

René kept to his heated pace, forcing Lang to press to his limit. The German driver tried to respond, but he was frustrated with the other cars on the course, two of which had been lapped over a dozen times.

By the sixty-fifth lap, René had widened his lead to one minute, twenty-seven seconds.

Then, to everyone's shock, Lang steered into the pits. He complained about the gearbox, as Rudi had done earlier, but the mechanics could find no obvious problem.

René swept past, already into his next lap.

Neubauer ordered Hermann to return to the race. Mercedes would not drop out. By the time the Silver Arrow was back on the course, René owned a three-minute lead. Undaunted by this long gap, Hermann attacked, clearly capable of overcoming whatever shifting trouble had vexed him so. On one lap soon after coming out of the pits, he nearly improved on the lap record set earlier by Rudi.

René slackened off a little on his speed. The race was his to lose now. Keep the Delahaye running. Into every corner and straight, stay true.

By lap seventy-four, Lang remained over a lap behind but had reduced his gap by eighteen seconds, to two minutes and forty-two seconds. Over the next six laps, he knocked off another ten seconds. Some among the crowd wondered if he might be able to stage a comeback. Would René and his Delahaye hold?

With each turn of the circuit, René drew back slightly on his pace. He read the lead times each time he passed the pits. He had set the stage to win, and now he needed only to perform his part to the end. Discipline and sharp focus. He felt welded to the Delahaye, he and the car operating with single-minded intention.

Lap eighty-five — two minutes, twenty-two seconds ahead.

Lap eighty-nine — two minutes, fourteen seconds.

Lap ninety-five — one minute, fifty-nine seconds.

Five more laps. René could almost taste the win now. Every time he came by the grandstands the crowd pulsated with noise and energy.

Lang continued to chip away at his lead, but it was too vast to make much of a difference. In the final lap, René sprang again. He mounted the hill and took the bend under the Pont Oscar. Out of the Lycée hairpin, Lang clung to the Delahaye's tail, albeit a lap behind, no doubt attempting to force a mistake from René, as his teammate had done earlier. René refused to oblige. Quickly enough, he entered the fast straight toward the final stretch as Hermann passed him in a useless attack.

The Delahaye engine rose to a deafening note as René came around to the finish. Faroux stood on the line, arms raised overhead. In his right hand he held the checkered flag, and as he waved it jubilantly, René Dreyfus crossed the finish. When he halted in the pits, the grandstand gates and barricades around the course appeared to break, and a tide of delirious people swirled around the Delahaye and its driver.

René lowered his goggles to his neck. A smile overtook the grim mien of determination that had been fixed on his face over the past three hours, eight minutes, and fifty-nine seconds. The race was over. He had set a new record time for the Pau Grand Prix. Better, by an indisputable margin of one minute and fifty-one seconds, he had triumphed over Mercedes.

A garland of flowers was placed on the Delahaye, and his crew handed René a fizzing bottle of champagne. Rising to the top of his seat, he drank to his victory — and theirs. A band struck up the "Marseillaise," his country's national anthem and rallying cry. The crowd sang the verses with uncommon emotion:

Arise, children of the homeland,
The day of glory has arrived!
Against us, tyranny's
Bloody standard is raised!

Epilogue

Torchlit parades wound through the streets of Berlin and Vienna. The votes from the plebiscite had been counted: "Ja" won over "Nein" by a margin of over ninety-nine to one. The Greater German Reich was now approved by the "people," and Hitler declared on the radio that the results "surpassed all my expectations . . . For me this hour is the proudest of my life." Outside the Austrian capital, huge swastika-shaped fires burned like warning signs to the world.

That same night in Pau, René celebrated his win in the Palais Beaumont, where a fine dinner was held in his honor, replete with champagne, speeches, and dancing under the twinkling chandeliers of the grand hall. When people congratulated him, René was humble, commenting that the stars of the "right day, the right driver, and the right car" had aligned in constellation.

The French press was far less reserved. In a string of articles, Charles Faroux could barely contain himself. "One cannot congratulate Delahaye enough for having beaten a rival as redoubtable as Mercedes-Benz . . . Could this be the dawn of a resurrection?" A *Paris Soir* editorial answered the question for him: "The success achieved in Pau by René Dreyfus and Delahaye thus marks the revival of the French sports industry." Newspaper headlines celebrated the same. "A Beautiful French Victory!" *Le Figaro* announced. "An Undisputed and Indisputable Victory!" exclaimed *L'Auto*.

British and American reporters also lauded the win. "There was something of a sensation at the Pau Grand Prix when the first race was held under the new Formula," wrote *Motor*. "Delahaye may with good reason be proud that their twelve-cylinder car, in the hands of a driver of first-rate caliber such as Dreyfus, beat the Mercedes fair and square."

The report in *Motorsport* read, "The first race run under the new formula was won by an unblown 4.5 liter car, in defiance of the forecasts of most people, and to the indescribable delight of thousands of Frenchmen, who stampeded across the road to congratulate their compatriot; cool, calm and unruffled René Dreyfus."

At the Hôtel de France, Kudorfer and the Daimler-Benz press team worked overtime to spin their interpretation of the ignoble loss. They tamped down the importance of Pau Grand Prix, once again calling it a "dress rehearsal" while highlighting how the W154 had achieved the fastest lap and had only lost because of its need to refuel.

German newspapers echoed their press releases: the Silver Arrow was "flawless," Pau was merely a "regional race," and in the important events ahead Mercedes was sure to come out on top. Still, the propaganda efforts could not deny that the victory by Delahaye was "incontestable." Privately, Neubauer and Rudi discounted the loss, but Mercedes driver Richard Seaman called it what it was: "an embarrassing defeat." Hans and Paula Stuck sent a telegram to René that simply said, "Phantastique."

Two weeks later, René won another Grand Prix — two in a row. On a six-mile road circuit west of Cork, in Ireland, he set a breakneck pace, all while being sprayed by hot oil from a gearbox leak. He beat Jean-Pierre Wimille and Prince Bira of Siam, who were driving Bugatti and Maserati single-seaters, respectively. No German cars competed that day.

It was an impressive start to the season for Écurie Bleue. "The first year of the new Grand Prix formula — two races so far," *Autocar* reported. "And both to France and both to Delahaye!"

Jean François promised that he would soon deliver the Delahaye 155, the single-seater that would exceed the performance of the 145. Lucy Schell hired a pair of Austrian engineers who had recently fled to France, because of their Jewish faith; they focused on boosting the power of the V12 that would be placed in the new car.

She also sent acerbic letters to the sporting press, asking how much in monthly advertisements she needed to take out to convince their editors to get their facts straight. "We always tend to imagine that Écurie Bleue is Delahaye," Lucy wrote. "This is not the case. While representing Delahaye, my stable is my absolute property and autonomous in every way." She asked, in the future, that mentions read something like, "Lucy Schell's Écurie Bleue."

All this came at the same time as the decision by the Fonds de Course to allocate its funds entirely to Talbot. Monsieur Charles was apoplectic, Lucy even more so. In response, she declared that her now-famous Delahayes would drop out of the French Grand Prix. "Please take note," she wrote the ACF, "that the only interested party able to go back on this decision is me." She had fought to the top of the Grand Prix game, she wanted the credit, and she wanted the support. Sadly, she received neither, but the accomplishment was hers nonetheless.

After Cork came the Tripoli Grand Prix. It took René one round of the Mellaha circuit to confirm what he already knew: his Delahaye stood no chance against the Germans on the long straights, where they would be able to exploit their nearly 200-horsepower advantage. That day Mercedes finished one-two-three, and René crossed the line far back in seventh place, his 145 almost aflame its engine was so hot.

Throughout the rest of the 1938 season, Mercedes dominated. By recruiting Nuvolari, Auto Union managed to win the Donington and Italian Grands Prix, but otherwise it was Mercedes who claimed the French, German, and Swiss races.

The Delahaye single-seater failed to live up to expectations, and René and Écurie Bleue never again challenged the overwhelming power of the supercharged engines of Mercedes and Auto Union. Nonetheless, his two Grand Prix wins (the only ones in a non-German car that season), a fourth-place finish at the Mille Miglia, and a first at La Turbie earned him the title of Racing Champion of France. Pau was his crowning achievement, and as Lucy intended, it was a blow against the invincibility of the Silver Arrows.

The symbolic importance of his victory was not lost on René, nor on the world. John Weitz never forgot listening to the race from London as a youth and feeling a tremendous sense of pride. For Weitz, who later became a successful American entrepreneur, amateur race car driver, and car designer in his own right, René was nothing short of a "divine avenger" for his people. Before the end of 1938, the Nazis would ravage Jewish communities throughout Germany, beginning with the infamous Kristallnacht. In a systematic orgy of terror, hundreds of synagogues were torched, thousands of businesses wrecked, and tens of thousands of Jews forced into concentration camps. Harbingers of hope were in sharp demand.

René was thirty-four at the start of the 1939 Grand Prix season, but thought he might have one or two more years left in him before he retired. He was already in talks with Monsieur Charles about coming on board at Delahaye as a sales director. Their many successes in competition had completely revitalized the firm, and production models based on the 145 had stolen the show at that year's Paris Salon.

His plan was to settle permanently in the French capital with Chou-Chou. Maurice, recently engaged, was there, and the two brothers looked forward to living and raising their families together. It was time to start thinking of such things.

Continued challenges with the Delahaye 155 hobbled the 1939 season for Écurie Bleue from the start. Lucy was distracted as well. While driving from Monaco to Paris before their last race the previous year, Laury got into an accident. He continued onward but checked into a hotel as soon as he reached Paris. He called Lucy to say he was okay, then went to sleep. In the middle of the night, he tried to reach out to something on the bedstand when he realized that he could not move. The accident had left him partially paralyzed.

The team did not compete until the French Grand Prix in July at Reims, where René finished a distant seventh. Days later, he left for the German Grand Prix, an ominous journey given that he carried in his suit pocket his call-up orders from the French army. War against Germany looked inevitable now.

René managed fourth place at the Nürburgring. At the dinner party afterward, a Mercedes representative raised a toast of "Vive la France!" to him for his performance in the race against much more powerful rivals. Since crossing the border into the Third Reich, René had been treated with every respect, but he could not overlook the escalating viciousness with which the Nazis treated Jewish people. Good decorum would have had René raise his glass and offer a "Vive L'Allemagne" in return. Instead, he remained seated, to an uncomfortable silence.

The German Grand Prix was his last race in a Delahaye. Never patient and still upset over the Fonds de Course decision, Lucy decided to switch her team to Italian Maseratis. At the Swiss Grand Prix on August 20, Mercedes swept the field one-two-three again. René was a lap behind in the straight-eight Italian car.

On September 1, the Nazis invaded Poland. Soon after, France declared war on Germany, and a maelstrom swept across the globe. Less than a month later, Lucy and Laury were being driven back to Paris from their Monaco villa when their chauffeur collided with a van. Both the Schells were seriously injured, and Laury died a month later. On leave from the French transport corps, Corporal René Dreyfus attended the funeral in Brunoy. A heartbroken Lucy remained in the hospital and missed the funeral of her beloved husband and co-driver. At the gravesite, René supported the Schells' two sons, Harry and Philippe, who were young men now.

In early January 1940, René lost his mother to illness. At the wake, Lucy informed René that she planned to send him to America to compete in the Indianapolis 500. He assured her that the army would never give him permission. In early May, his commanding officer called him into headquarters. Orders from the High Command were that René was to represent France at the American race. Lucy had got her way again.

He tried to convince Chou-Chou to join him, but she said that she wanted to stay in France to look after her mother. There was a shadow over their marriage now, but René could not quite discern its cause.

While he and Harry Schell, a rambunctious nineteen-year-old, were on their way across the Atlantic, Germany stormed across the border into France. There was nothing René could do but read the newspapers and shudder at the horrifying reports. His first days in America were exciting but bittersweet given the events at home in France. He finished tenth at the Indianapolis 500, in a Maserati, his race strategy foiled by rules that bewildered him, including a restriction against passing other cars while it was raining.

The situation in Europe quickly deteriorated. René wanted to go back, but in telegram after telegram Lucy and Maurice urged him to stay in the United States. Paris fell soon after, and then France surrendered to Hitler. Discharged from the army, adrift in America, unable to speak English, his bank accounts frozen, René tried to forge a life for himself. He found a meager basement apartment in New York, and while living there, he learned his wife was having an affair with an official in the Fascist Vichy government. When Chou-Chou petitioned for divorce, her legal grounds were that René was Jewish. Then all word from France ceased. He did not know if his family was alive or dead.

Eventually René opened a restaurant in New Jersey with a Niçois naval officer of some means acting as his partner. René knew a thing or two about fine food and was a born connector, and their business flourished, particularly among wealthy European expatriates. René managed to get by speaking mostly French and Italian, which only made him more homesick.

After the attack at Pearl Harbor, René enlisted in the US Army. He underwent basic training at Fort Dix, South Carolina, and learned English at last. He also became an American citizen. Finally, in the spring of 1943, Staff Sergeant Dreyfus left for Europe and the war. His transport ship's newspaper made a big deal about how they had a celebrity on board: "Dreyfus was to the French people what Babe Ruth was to the Americans." After a brief stay in Morocco, he participated in the September 1943 Allied invasion of mainland Italy. Commanding a transport company of two hundred trucks, René endured ferocious German shelling at the landing on the beaches of Salerno. The horrors he witnessed stayed with him for the rest of his life.

After the battle, a newly promoted René moved on to Naples, then Rome, where he served as a translator for captured Italian soldiers. As soon as the south of France was liberated, in August 1944, he caught a military plane to Nice, desperate to know if his brother and sister had survived the Vichy regime. They had not seen each other since their mother's funeral, nor had they communicated in almost four years.

He went to his sister Suzanne's last known address, and as he approached the door, he said later, he could already sense her presence. They soon were weeping in an embrace, so overcome with emotion that words failed them. Suzanne's husband ran off to find Maurice. At the news that his younger brother was not only alive but in the city that very moment, Maurice ran out of his apartment without his pants on.

When the three siblings reunited, stories of their years apart spilled out like spools of thread. Maurice and Suzanne had both participated in the French Resistance. Betrayed, Maurice had spent months in hiding, evading the Gestapo.

In the early hours of the morning, the three spoke of the future. America, René pitched, was the place for them.

When war broke out, Rudi and Baby Caracciola retreated to their chalet in Switzerland. Although fit enough before to drive at the highest level of the Grand Prix, he bowed out of any service to Germany on the stated account of his disability. To supplement the meager food available in stores, he and Baby dug up their flower gardens to plant potatoes, corn, and string beans. They kept chickens as well. While battles raged across the world, they lived like hermits, rarely leaving Casa Scania.

In mid-1941, Rudi ventured to Stuttgart. He had one purpose: to obtain a 1.5-liter Silver Arrow that he had driven to a second-place finish in the all-*voiturette* 1939 Tripoli Grand Prix. Once the war was over, Rudi wanted to get straight back into racing, and this was the perfect car for him. He could think of nothing else.

Kissel agreed, but only if he could get government approval for the export license, a hard ask at that point. Rudi returned to Switzerland and waited. Meanwhile, fearing the *voiturettes* and other Silver Arrows were at risk from bombing raids, Mercedes cemented them inside air-raid shelters or secured them in barns in the German countryside.

Throughout 1942, while his former teammates, like Manfred von Brauchitsch and Hermann Lang were participating in the war effort (albeit far from the front lines), Rudi stayed in Switzerland, drawing his pension. This finally caught the attention of the Nazi high brass, who did not care for their top race car driver sitting out the fight in a neutral country. There were even rumors that Rudi was speaking ill of Germany's chances.

Alfred Neubauer defended Rudi, promising that he was loyal to his country but that his injuries made him unable to serve. The regime cut off his monthly stipend, and Rudi was warned by his former team manager to stay in Switzerland.

During the war, Daimler-Benz continued to be a leading armament producer for the Reich. Its assembly lines made airplane engines, tanks, heavy trucks, armored vehicles, and all the spare parts the Reich de-

manded. The company built massive plants in Germany and appropri-
ated factories in Nazi-occupied countries. Needing workers by the tens
of thousands for these factories, its Stuttgart leaders did not hesitate to
force occupied populations, prisoners of war, and concentration-camp
inmates into slave labor under draconian conditions. All the while, profits soared.

To get all these vehicles on the road, the German army required drivers and mechanics. Thanks to the flood of recruitment during the heyday of the Silver Arrows, the NSKK provided a force of 187,000 trained
drivers. Although many joked that NSKK stood for Nur Säufer, keine
Kämpfer (Only Drunkards, Not Fighters), the organization was very
effective in contributing manpower to the Wehrmacht's motorized in-
fantry divisions. For this reason Hitler awarded Hühnlein, on his death
from natural causes in 1942, the German Order, the Third Reich's highest honor.

NSKK troops proved essential in building the fortified Siegfried Line
of defenses, operating transport brigades, and executing blitzkrieg operations across Europe. Their motorcycle troops were the self-declared
"spearhead of the modern war." NSKK ranks also played an active role
in the persecution of the Jews, starting with erecting roadblocks and
bringing in stormtroopers during Kristallnacht. It was NSKK trucks
that carried out the deportations and the ghettoization of vast numbers
of Polish Jews. Throughout the Holocaust, they perpetrated numerous
atrocities, including the shooting of Jews, especially in Russia and the
Ukraine.

At the end of the war, after years of relentless Allied aerial attacks,
Germany was a hollowed-out shell. Stuttgart had been a high-value
target — especially the Daimler-Benz plants at Untertürkheim — and
heavy bombing had left it a forest of bent iron columns and heaps of
rubble.

As for their Silver Arrows, most of them were grabbed by the Rus-
sians, to be studied at the Soviet Union's technical schools. They showed
no interest, however, in the Mercedes *voiturette* that Rudi had been seeking obsessively. After Germany's surrender, while most struggled to endure the wasteland that was Europe, he continued his quest to obtain
one so that he could drive competitively again. At last he succeeded and
soon after was invited to participate in the 1946 Indianapolis 500.

After all his efforts, export restrictions forbade the transport of the cars to the United States, but Rudi traveled to Indianapolis anyway and was offered a Thorne Special to drive. During the qualifying rounds of the race, an object, ostensibly a rock churned up on the track, struck him by the temple. He was momentarily stunned and crashed the car. A severe concussion left him more or less unconscious for almost a week, and it took years of recovery in Switzerland before he could drive again.

During that time, Mercedes returned to racing, first with sports cars. Neubauer again led the team. He brought in a new crop of drivers to join Lang, including Karl Kling and Juan Manuel Fangio. Rudolf Uhlenhaut designed the cars, starting with the Mercedes 300 SL, a design that confirmed his standing as an engineering legend.

In 1952, Neubauer enlisted Rudi to represent Mercedes at the Mille Miglia. Now fifty-one, Rudi finished an impressive fourth. A few weeks later, he competed in another sports-car race at Berne. Coming out of a turn, his brakes locked up and he smashed into a tree. This time his left leg bore the brunt of the impact. "Rudi, oh Rudi," Baby murmured when she saw him in the Red Cross tent afterward. He spent the next year in hospitals, bound to his bed, and often in excruciating pain. He never competed again.

In 1955, the year Neubauer retired after renewing the Silver Arrows' dominance in the Grand Prix, Mercedes gave Rudi a job pitching their cars to American and British troops in Europe. Although successful in this new venture, he was drinking heavily, and he died from cirrhosis of the liver in 1959. The firm gave him a hero's funeral that whitewashed away his role as a standard-bearer of the Third Reich—just as his own autobiography, published after the war, had done. No doubt Rudi Caracciola was one of the twentieth century's finest Grand Prix drivers, but most people chose to forget the devil's bargain he made to achieve this dream.

Rudi's former teammate and longtime rival Tazio Nuvolari also returned to the circuit when peace reigned again in Europe. He and his wife, Carolina, had lost one son to heart trouble in 1937. Then their other son, and last remaining child, died of a kidney disorder in 1946. Afterward he bemoaned, "There is no longer any object in life for me. If I do not return to racing I am finished. Cars are the only thing I have left. I need them; I must travel round the world again. I cannot forget, but I must alleviate my pain."

The Flying Mantuan, now fifty-four years old, managed some top finishes over the next several seasons, often piloting a new Ferrari-made car, but age and his many injuries caught up with his long-battered body. He never officially retired, but he drove his last race in 1950 at the age of fifty-seven. Two years later, a stroke left him bedridden and partially paralyzed before he passed away in 1953. In Mantua, a mile-long procession followed his coffin to the cemetery. Above the family crypt in which he was buried were these words: CORRERAI PIU FORTE PER LE VIE DEL CIELO (HE WILL RACE FASTER THROUGH THE STREETS OF HEAVEN).

As for Louis Chiron, he never married. He continued to compete in Grand Prix races through his fifty-eighth birthday. After finally retiring, he became the charming, always loquacious master of ceremonies at the Monaco Grand Prix, a race he had helped launch.

During the occupation of Paris, Charles Weiffenbach labored in "constructive non-cooperation" to keep the Delahaye factory from producing much that might aid the Nazi war machine. He was not alone within the French racing community in his resistance efforts. Most notably, drivers Robert Benoist, William Grover-Williams, and Jean-Pierre Wimille created a network of sabotage cells with the British Special Operations Executive (the famous "Ministry of Ungentlemanly Warfare") to fight back against the Paris occupation. Tragically, the Nazis captured, tortured, and killed Benoist and Grover-Williams.

In September 1945, barely a year after the Allies freed Paris, the revival of European racing began on a 1.7-kilometer circuit in the Bois de Boulogne, and the initial race was fittingly named the Coupe Robert Benoist.

After the war, Monsieur Charles led another resuscitation at the rue du Banquier, this time without Jean François, who had died of a lung disease in 1944. Now in his mid-seventies, Weiffenbach remained untiring, concentrating on much-needed heavy trucks and variations of their tried-and-trusted Delahaye 135. Coach-builders like Henri Chapron, Letourneur & Marchand, Figoni & Falaschi, and Jacques Saoutchik created splendid design classics with these cars; as one automobile historian noted, Delahaye was "the undisputed star of the salons."

Nevertheless, the company struggled to survive in a crippled Europe and a French state bent toward socialism. By the mid-1950s, Monsieur Charles and Delahaye's owners were forced to merge with the French firm Hotchkiss. When this combined entity was sold, mere months later, the Delahaye marque disappeared into obscurity.

Sometime after the Nazi occupation of France, Lucy left for the United States with her son Philippe. They remained there until the armistice, and then she and her two sons returned to Monaco. She never competed in motorsport again. The American speed queen, one of the preeminent Monte Carlo Rallyers for a decade and the first—and only —woman to own and lead a major Grand Prix racing team, faded from memory. In 1952, she died and was buried beside her beloved Laury in the cemetery in Brunoy.

Twice a week, long before the sun came up, René would drive his Peugeot down to the Fulton Fish Market in downtown Manhattan. He made sure to be there at 5:00 a.m.—not so early that he was sold the previous day's catch, not so late that he missed the best selections of the day: scallops, mussels, striped bass, soft-shell crabs. He would load the fresh fish and seafood into the trunk of his car and bring it to 18 East Forty-Ninth Street, a stone's throw from Rockefeller Center, where Le Chanteclair, the restaurant he had opened with Maurice in 1953, was located.

Typically, by the time René arrived from the fish market, his brother would already be there. Together, they bustled about the quiet of the restaurant, preparing for the day's service with the chef and staff. Then the two brothers would return to Forest Hills, Queens, where they lived, a few blocks apart, for a few hours' rest before the restaurant opened.

By early evening, Le Chanteclair was astir with conversation, laughter, drinks, and a steady delivery of fine French classic dishes. Elegantly dressed in suits and ties, René and Maurice alternated in the role of maître d', greeting and chatting with their patrons and managing the front and back of the house. Their sister Suzanne handled the cash and reservations. Six days a week, often for fourteen hours a day, the siblings ran the place together.

For more than a quarter century, Le Chanteclair was a favorite on the New York scene and the gathering place of the city's car and racing enthusiasts. The wall opposite the long bar was covered with photos of top international drivers—the old guard of Nuvolari, Chiron, Wimille, and even Caracciola, as well as the new generation: Phil Hill, Stirling Moss, A. J. Foyt, Jackie Stewart, and others. Some of this new generation were regular patrons, and René became a kind of godfather to many of them.

Past the bar was the dining room, covered with murals of French scenes such as the Place de la Concorde, and on any given night diners passed the time until their food arrived in spotting celebrities, from the world of racing and elsewhere. Walter Cronkite, Neil Armstrong, Edith Piaf, Billie Jean King, and William Faulkner were commonly seen. They came for the food, the fine wine cellar, and, most of all, the entertaining company that the brothers Dreyfus cultivated.

Le Chanteclair allowed René to remain part of the racing world, and his passion for the sport never waned. He captained a few racing teams and drove honorary laps at some Grand Prix events, such as the fiftieth anniversary of his win at Monaco in 1980. On a whim, he ran Le Mans once more in 1952, but his Ferrari broke down midrace. He told one interviewer soon after his restaurant closed in 1979, "I loved racing. I love it, and I will love it until I die."

As the title of his memoir, *My Two Lives,* indicates, René enjoyed two long careers: the first as a top French Grand Prix driver, the second as a renowned French American restaurateur. This did not even include his years in the US Army, which he considered his proudest achievement. Throughout his life, he remained agnostic—and downplayed the significance of his heritage in his Pau victory. "I am listed in the *Encyclopedia of Jews in Sports,*" he modestly wrote in his memoir. "Were there one, I would qualify for the Catholic sport encyclopedia too."

In 1993, René was diagnosed with an aortic aneurysm. At eighty-eight years of age, he had an even chance of surviving open-heart surgery. Unsurprisingly to anyone who knew him, he chose to take the risk. The night before the operation, he spoke by telephone to a good friend in France. They relished memories of his driving years, most pointedly his 1938 win over the German Silver Arrows. At the end of their conversation, René wept and then said hopefully, "I've lived so many things . . . I've brushed with death many times. Maybe I'll make it out once again."

It was not to be. Shortly before the surgery was finished, René passed away. In accordance with his final wishes, friends spread some of his ashes at the places of his finest triumphs, including Montlhéry and Pau.

When German forces took over Paris in 1940, they seized the records of the Automobile Club de France. There were rumors that Hitler also sent men to find and destroy the Delahaye 145 that had defeated his Silver Arrows at Pau. The initial fates of the four Écurie Bleue 145s owned by Lucy Schell remain murky, but she probably sold them off with the aid of Charles Weiffenbach, who took possession of them before the Nazi invasion. Two were known to have been kept together. Coach-builder Henri Chapron disassembled them and scattered their parts deliberately about his shop amid other car parts. The other two were hidden more remotely, in either barns or caves far from the capital.

After VE Day, the movements of the 145s became better documented. Like practiced detectives, a pair of Delahaye historians, Richard Adatto and André Vaucourt, tracked down and assembled reams of information on the fate of each 145. The records from the ACF, however, were never found.

Chapron rebodied the pair of 145s that stayed together throughout the war. They became instant classics of high art-deco design and, over several decades, were traded among a host of enthusiasts until they ended up in the automotive museum of Peter Mullin in Oxnard, California. Mullin, who made his fortune in the insurance business, has a particular passion for prewar French cars, which he considers the "apex of the automobile world in terms of engineering genius, performance, and beauty." He owns a priceless collection of Delahayes, Citroëns, Delages, Bugattis, Hispano-Suizas, and Talbot-Lagos.

The other two 145s had more complicated journeys. One received a postwar convertible roadster body from French coach-builder Franay and was included in a "most beautiful cars in the world" exhibition. Eventually, a new owner, Philippe Charbonneaux, placed a sports-car body on the 145 that harkened back to its 1938 Mille Miglia design. At the fiftieth anniversary of the Million Franc prize, in 1987, René Dreyfus had the joy of taking this restored Delahaye back onto the Montlhéry track and driving it at nerve-rattling speeds. This car was bought by another American, industrial designer Sam Mann, who placed it in his collection of "rolling sculpture" in Englewood, New Jersey.

The last of the four Delahaye 145s participated in some postwar races. Then, for a long while, it sat a half-bodied derelict under the Montlhéry track before being moved to a French chateau, one hour west of Paris. Peter Mullin and a partner purchased this Delahaye in 1987. After an expensive restoration in England, Mullin included it in his collection. Like Sam Mann, Mullin has presented the Delahaye at Concours d'Élegance events and has also participated in several classic car races.

ADATTO ARCHIVES

The Delahaye 145 found at Château du Gérier,
owned by collector Serge Pozzoli

A debate has long simmered between the two American collectors over who owns the Delahaye 145 in which René claimed the Million Franc prize and won the Pau Grand Prix. It is a question that could mean the difference between millions of dollars as well as a fair bit of value in terms of pride. Each has a homologation expert (Adatto and Vaucourt) on his side who has studied the chassis and engine numbers, plumbed the French archives, combed through contemporaneous photographs, and examined every inch of the cars for clues of provenance. The Chapron-bodied Delahayes have largely been ruled out, but that leaves a duel between two.

A race car is not a static object like a painting. Throughout its racing days, its engine is rebuilt and rebuilt, its body altered, its parts replaced. Since they retired from the circuit, the 145s have seen their engines overhauled and their bodies swapped many times over. In restoration, careful attention was paid to mirroring the exact details of the original car, but the fact is that many of the parts needed to be fabricated anew. For the sake of fairness and equanimity, one might imagine that both restored 145s contain some parts from the race car that Dreyfus drove.

As with religious relics, authenticity (or proof thereof) is often beside the point to most people. It's the emotion invoked by the object that is paramount. One cannot walk around the sleek bodies of either the Mullin 145 or the Mann 145 — or listen to the growls of their V12s — and fail to be moved. Both bring to mind a great struggle in which David conquered Goliath, and, for a moment, there were heroes again in the world.

Acknowledgments

As much as short and sweet is probably the best way to go in delivering acknowledgments, there is simply no way to do so with *Faster*. I am indebted to many people around the globe who aided in delivering this history.

First and foremost, Richard Adatto. He is a Delahaye and French coach-builder historian, and since we were both living in Seattle at the time, he was one of the first people I reached out to, hoping he might have an insight or two. Our brief dinner turned into a long collaboration, which made this book richer far beyond measure. Richard opened up his extensive archives to me, offered introductions, provided research guidance, shepherded me through the Mullin Museum, and gave me a taste of what it was to drive the "Million Franc" Delahaye. He is a prince of a man.

In France, Jonathan Dupriez was my indefatigable researcher. He opened many doors for me and tracked down some obscure, but key, pieces of information about Lucy Schell and her family. Together, we toured some of the great Grand Prix racecourses of the 1930s. In Germany, Almut Schoenfeld, Marco Pontoni, and Maren Michel gave incredible assistance, helping to uncover the history of the Mercedes-Benz Silver Arrows and their links to the Third Reich. Thank you also to Almut for deciphering some texts for me that would have otherwise remained a mystery. Finally, in the United States, Claire Barrett, Nicole

Diehm, and Tatiana Castro hunted down numerous texts, magazine articles, and archives on my behalf. Their tireless efforts are greatly appreciated. My thanks also to André Vaucourt, a passionate, deeply knowledgeable expert on Delahaye who provided countless specifics on the French automotive firm from his own vast research.

Of the four Delahaye 145s built, collectors Peter Mullin and Sam Mann have cornered the market. They both met with me personally, shared their love of classic French cars, and generously gave me the opportunity to hear the roar of the V12 engine. Their passion for these race cars was infectious. Thanks also to Fred Simeone, who has his own treasures from the 1930s, in Philadelphia.

A big shout-out to Evelyne Dreyfus for providing me with a wonderful lunch and fond recollections of her father Maurice, her uncle René, and the whole Dreyfus family. She also read an early version of the manuscript to make sure I had the story straight. Thank you for the many kindnesses, Evelyne. I also owe much to the pioneering research of Beverly Rae Kimes, co-author of René's memoir, a fine text in its own right. Her widowed husband Jim Cox shared the original audio recordings of Beverly's interviews with René. What a treasure. Jon Bill, curator of the Auburn Cord Duesenberg Automobile Museum, which holds many of Beverly's archives, gave much assistance to plumb her original research as well.

Many, many folks illuminated for me the delightfully colorful world of the 1930s Grand Prix, shedding light on the masterpieces by Bugatti, Maserati, Mercedes-Benz, Auto Union, and Alfa Romeo and the characters who manufactured and competed in them. Thank you notably to Hervé Charbonneaux, Maurice Louche, Michel Ribet, Serge Bellu, Alfred Wurmser, Christian Schann, the late Pierre Darmendrail, Lydie Chiron, and Karl Ludvigsen. Special gratitude goes to Leif Snellman, who has probably collected more research into European racing in the 1930s than any other individual. His website devoted to "The Golden Era of Grand Prix Racing" is a must-have resource for any researcher of this period. Time and again, I drew from its pages.

As any historian knows, the dedicated efforts of librarians and archivists make possible our endeavors to bring stories of the past alive to readers. *Faster* is no exception. Particular thanks to: Mark Vargas, director of the REVS Institute Library; Pierre Fassone, local memory of

Châtel-Guyon; Patricia Bourgeix for her tour of Montlhéry; the staff at the Daimler-Benz Archive, chiefly Gerhard Heidbrink, Wolfgang Rabus, Michael Jung, and Silvie Kiefer; the staff at the Automobile Club de Basco-Béarnais in Pau, Nathalie Verdino at the Automobile Club de Monaco, and André Gervais, archivist at the Automobile Club de Nice.

Two noted automotive enthusiasts, classic car collectors, and generally knowledgeable fellows, Stephen Curtis and Flavien Marçais, were keenly helpful in guiding me to research, sharing their own, and, perhaps most important, combing through the first drafts to ensure that I did not mistake a carburetor for a camshaft, not to mention pointing out where I might have gotten the history wrong. Any errors, mistakes, or misunderstandings that remain are my burden alone to carry. Thank you, fine gentlemen. Carl Bartoli, my lovely father-in-law and engineer-on-call, was also helpful in making heads-and-tails of certain automotive systems, while also aiding with some important Italian-to-English translations.

Once again, I express my eternal gratitude to my publishing team at Houghton Mifflin Harcourt for giving me the opportunity to enjoy my passion for great narrative history. Susan Canavan first championed this project, and then Alex Littlefield stepped in to provide an incredible edit that made this book far better than it ever would have been in the first place. Masterful work. Thanks to Olivia Bartz for assisting me with the photos, Lisa Glover for steering this through production, and Lissa Warren for drumming up publicity to spread the word. As ever, my first line of defense, Liz Hudson, helped shape every paragraph and page. I'm lucky she still puts up with me after these many years. Finally, on my publishing team, nothing gets done without my agent Eric Lupfer, who always has a wise word and a keen eye for what needs to get done. Appreciations to Anna Deroy and Andy Galker for the film work.

Sam Walker and Christy Fletcher, longtime New York friends, set me on my journey to uncover the story of René Dreyfus, Lucy Schell, and the Delahaye 145. You two are the best, and that French chateau awaits your arrival.

Finally, as always, to my wife, Diane, and two spirited daughters. I only wish life would go slower.

Notes

Abbreviations

ACBB — Automobile Club Basco-Béarnais, France
ACF — Automobile Club de France, France
MMA — Mullin Museum Archives, United States
DBA — Daimler-Benz Archive, Germany
REVS — Revs Institute, United States
PPRA — Personal Papers of Richard Adatto
PPDF — Personal Papers of the Dreyfus Family
PPBK — Personal Papers of Beverly Kimes, Cord Duesenberg Library
PPML — Personal Papers of Maurice Louche

Author's Note

page

xii *leader of the new Third Reich made:* Rosemann, *Um Kilometer und SeKunden,* pp. 1–2; Klemperer, *The Language of the Third Reich,* p. 4.
"A Mercedes-Benz victory": Bretz, *Mannschaft und Meisterschaft,* p. 14.

xiii *"crooked-leg racehorse":* Hildenbrand, preface. I owe a debt of gratitude to Laura Hildenbrand, author of the classic narrative-nonfiction history *Seabiscuit.* Her structuring of the opening — and introduction of her main characters—served as a template for structuring *Faster*. If imitation is indeed the sincerest form of flattery, then I hope Laura is blushing.

xvii *"this life of fearful joys"*: Ferrari, *My Terrible Joys*, p. 13.

Prologue

xix *The beast, long lurking*: Liddell Hart, *History of the Second World War*, pp. 65–75; Nicholas, *The Rape of Europa*, p. 81.

border between France and Germany: Shirer, *The Rise and Fall of the Third Reich*, p. 723.

By May 15, the French prime minister: Lottman, *The Fall of Paris*, pp. 46–47.

xx *"a cruel machine"*: Walter, *Paris Under the Occupation*, p. 13.

At the news: Blumenson, *The Vilde Affair*, pp. 26–27.

"With bicycles and bundles": Lottman, *The Fall of Paris*, p. 143.

Fearing an invasion: Nicholas, *The Rape of Europa*, pp. 50–88; Rosbottom, *When Paris Went Dark*, pp. 7, 38; Lottman, *The Fall of Paris*, pp. 7, 64–65.

In the Delahaye factory: Dorizon, Peigney, and Dauliac, *Delahaye*, p. 60; undated news clip, PPBK; Charles Fleming, "A Classic Car Mystery," *Los Angeles Times*, November 2, 2015; Mullin Automotive Museum, "Delahaye Type 145" (brochure), MMA; notes of André Vaucourt, in the author's possession; Lew Gotthainer, letter to William Smith, December 6, 1971, PPRA. The fate of the Delahaye 145s during the war is one rife with contradicting stories and apocryphal legend, and the details of which of the four cars raced where is debated among Delahaye experts to this day.

In late May: Rosbottom, *When Paris Went Dark*, p. 51.

xxi *While the police*: Lottman, *The Fall of Paris*, p. 165.

"a swarm of bees": Ibid., p. 174.

Two days later: Liddell Hart, *History of the Second World War*, pp. 84–85; Shirer, *The Rise and Fall of the Third Reich*, pp. 721–23.

Finally, on June 14: Blumenson, *The Vilde Affair*, pp. 32–34; Mitchell, *Nazi Paris*, p. 3; Lottman, *The Fall of Paris*, pp. 340–70.

"DEUTSCHLAND SIEGT AN ALLEN FRONTEN": Mitchell, *Nazi Paris*, p. 13.

Trucks fitted with loudspeakers: Rosbottom, *When Paris Went Dark*, p. 50.

xxii *"Is the last word said?"*: Blumenson, *The Vilde Affair*, p. 38.

Marshal Philippe Pétain: Shirer, *The Rise and Fall of the Third Reich*, pp. 741–42.

"They knew where": Rosbottom, *When Paris Went Dark*, p. 64.

xxiii *Founded in 1895: ACF Bulletin Officiel,* June 1935; Lemerie and Piat, *Histoire de l'Automobile Club de France,* pp. 17–30.

One day early in the occupation: Richard Adatto, interview with the author, Seattle, 2017; Lemerie and Piat, *Histoire de l'Automobile Club de France,* p. 45. The story of the theft of the ACF archives derives from an interview with Delahaye historian Richard Adatto. Years ago, he met with the ACF librarian profiled in this scene. The librarian recounted to Richard the tale of the theft. Confirmation of the disappearance of the archives comes from Lemerie and Piat's book, among others.

The club's mahogany-paneled bars: Lemerie and Piat, *Histoire de l'Automobile Club de France,* pp. 35–43.

"Bring me all the race files": Adatto, interview with the author.

1. The Look

3 *May 19, 1932:* Brauchitsch, *Ohne Kampf Kein Siege,* pp. 10–13.

Of average height: "Getting Younger as He Ages," undated news clip, PPBK; Cholmondeley-Tapper, *Amateur Racing Driver,* p. 51.

"The Look": Sports Car Graphic, September 1974.

René drew up a chair: Neubauer, *Speed Was My Life,* p. 38.

4 *Many likened Berlin:* Weitz, *Weimar Germany,* pp. 161–65.

"To tread its streets": François-Poncet, *The Fateful Years,* p. 11.

Although Jewish by descent: Dreyfus and Kimes, *My Two Lives,* p. 54.

"The victor is sitting": Brauchitsch, *Ohne Kampf Kein Siege,* pp. 10–13.

5 *René circled: L'Automobile sur la Côte d'Azur,* May 1926; Evelyne Dreyfus, interview with the author, France, 2018.

The young Niçois: "René Dreyfus, le Driver Gentleman," *Auto Passion,* n.d., Dreyfus Magazine Scrapbook, PPDF.

René turned the fuel-line: Ted West, in "Rising to Greatness" (*Road and Track,* March 1987), interviews Dreyfus about his 1926 hill climb career. It's a superb article with great details, not only on what Dreyfus thought and felt, but also on the experiences (sounds, sights, smells) of that day. This scene, including the *"blaaattt"* and *"rrrraapppp"* of the Bugatti engine, draws on this article. Contemporaneous photos of René Dreyfus, PPDF, DBA.

"A racing Bugatti engine": Purdy, *The Kings of the Road,* p. 23.

6 *Considered the father: L'Automobile sur la Côte d'Azur,* January 1932; *Motor Sport,* December 1951; Jellinek-Mercedes, *My Father, Mr. Mercedes,* p. 88.

alone against the clock: René Dreyfus, interview with Jean Paul Caron, 1973, PPRA.

7 *five minutes, 26.4 seconds: L'Automobile sur la Côte d'Azur,* May 1926.
Rousing congratulations: Road and Track, March 1987.

8 *As a boy:* Dreyfus and Kimes, *My Two Lives,* pp. 1–2.
On occasion, his father: "Maurice Dreyfus — The Other Half," undated news clip, Personal Papers of Karl Ludvigsen, REVS.
War against Germany: Dreyfus and Kimes, *My Two Lives,* pp. 2–4.

9 *René felt unmoored:* "René Dreyfus, le Driver Gentleman," *Auto Passion,* n.d., Dreyfus Magazine Scrapbook, PPDF.
"sporting young bloods": Dreyfus and Kimes, *My Two Lives,* p. 4.
René entered: "Maurice Dreyfus — The Other Half," undated, uncited article in Personal Papers of Karl Ludvigsen, REVS; *Autosport,* March 11, 1955; *Sports Car Guide,* September 1959.

10 *By one calculation:* Purdy, *The Kings of the Road,* p. 21.
Before the Great War: Alfred Wurmser, telephone interview with Jonathan Dupriez, 2018; Christian Schann, interview with Jonathan Dupriez, 2018.
"little box of speed": King, *The Brescia Bugatti,* p. 30.
"marvel in the matter": Ibid., pp. 64, 7.
"jar the back teeth": Purdy, *The Kings of the Road,* p. 15.
"I suspect that": King, *The Brescia Bugatti,* p. 97.
"nimble Brescia": Dreyfus and Kimes, *My Two Lives,* p. 6.

11 *The Dreyfus family:* "René Dreyfus, le Driver Gentleman," *Auto Passion,* n.d., Dreyfus Magazine Scrapbook, PPDF.
every young man in town: René Dreyfus, interview with Maurice Louche, PPML; Court, *A History of Grand Prix Motor Racing,* pp. 56–58; Daley, *Cars at Speed,* p. 142.
After his grandson's 1926 triumph: Dreyfus and Kimes, *My Two Lives,* p. 6.
Over the next few years: Dreyfus race record, PPBK.
A half-dozen years: Road and Track, July 1962; *Motor,* April 17, 1934; *Motor,* August 22, 1933; Rao, *Rudolf Caracciola,* pp. 441–42.
"poised, handling his car": Nolan, *Men of Thunder,* p. 126.
A series of win-place-or-shows: Dreyfus and Kimes, *My Two Lives,* pp. 6–8; *Sports Car Guide,* September 1959.

13 *"child with a new and better toy":* Dreyfus and Kimes, *My Two Lives,* p. 16.
"Come on, we have some work": Auto Age, August 1956; *Sport Review's Mo-*

torspeed, (no month) 1953, Dreyfus Magazine Scrapbook, PPDF; *Pur Sang,* Spring 1980; Moity, *Grand Prix de Monaco,* 1930/4–8.

14   *the morning of the race: Sport Review's Motorspeed,* (no month) 1953.
  Monaco's small population: Autocar, April 11, 1930.
  "Think you've got a chance?": Sport Review's Motorspeed, (no month) 1953.

15   *In his spotless overalls: Sports Car Guide,* September 1959.
  He and the other drivers: Automobile Quarterly, Summer 1967.
  one last check of his car: This account of the overall race is drawn from a number of sources, most with first-person interviews of Dreyfus. Of particular note, I am indebted to Leif Snellman and his site on Grand Prix racing in the 1930s, where he has assembled a race-by-race account with meticulous care, drawing on a range of contemporaneous newspapers and magazines. See the 1930 Monaco Grand Prix entry at *The Golden Era of Grand Prix Racing* (website), http://www.kolumbus.fi/leif.snellman/gp3002.htm#9. Other sources include: *L'Intransigent,* April 2–10; *L'Auto,* April 2–10; *Auto Age,* August 1956; Moity, *Grand Prix de Monaco,* 1930/4–8; *Sports Car Graphic* (Monaco Public Relations), PPRA; *Road and Track,* September 1980; *Autocar,* April 11, 1930; *Motor Sport,* May 1930.
  "Don't force too much": Sport Review's Motorspeed, (no month) 1953. Notable sources for dialogue and special details are noted separately.
  At the line: Chakrabongse, *Road Start Hat Trick,* p. 38; Cohin, *Historique de la Course Automobile,* p. 112.
  "multicolored serpent": Court, *A History of Grand Prix Motor Racing,* p. 189.

17   *At the top:* Ibid., pp. 9–11; Daley, *Cars at Speed,* pp. 50–62; Hodges, *The Monaco Grand Prix,* pp. 8–13; Chakrabongse, *Road Start Hat Trick,* pp. 33–35.
  "dive into a dull, stone-side ravine": Lyndon, *Grand Prix,* p. 10.
  "What a crescendo": Motor, April 1930.

18   *"the Devil's course":* Brauchitsch, *Ohne Kampf Kein Siege,* p. 92.
  He drove the Bugatti: Dreyfus, interview with Caron, 1973; Lyndon, *Grand Prix,* p. 11.
  "He's there! He's there!": Pur Sang, Spring 1980.

19   *"Could Dreyfus do it?": Motor Sport,* May 1930.
  "Be very careful": Road and Track, September 1930.
  Louis was in a rage: Tragatsch, *Die Grossen Rennjahre 1919–1939,* pp. 119–20.
  Many celebrations: Dreyfus and Kimes, *My Two Lives,* p. 22.

"In magnificent style": L'Auto, April 7, 1940.

Between his first-place prize: Moity, *Grand Prix de Monaco,* 1930/4–8.

20   *The snubbing:* Draft of a preface by René Dreyfus for Franco Zagari's book on Maserati, November 1978, PPBK. This is a remarkable, extended version of the final preface that Dreyfus delivered for Orsini and Zagari, *Maserati.*

He returned heartbroken to Nice: Manuscript draft of Dreyfus memoir, PPBK; Ludvigsen, *Classic Grand Prix Cars,* p. 99.

Represented by the trident symbol: Ludvigsen, *Classic Grand Prix Cars,* p. 99.

"jumped about on its own suspension": Ibid., p. 99.

at the ramshackle Maserati factory: Sport Review's Motorspeed, (no month) 1953, Dreyfus Magazine Scrapbook, PPDF.

21   *"What do you think":* Ibid.

a stultifyingly hot: Brauchitsch, *Ohne Kampf Kein Siege,* p. 7.

René took off: L'Auto, May 23, 1932; *Autocar,* May 27, 1932; *Motor,* May 24, 1932; see the 1932 Avusrennen entry at *The Golden Era of Grand Prix Racing* (website), http://www.kolumbus.fi/leif.snellman/gp3206.htm#26.

22   *Early on, René set:* Old Cars, January 13, 1976.

"It's over": Dreyfus's draft preface for Zagari's book on Maserati.

Later at the hotel: Orsini and Zagari, *Maserati,* pp. 8–9; *Motor,* April 7, 1931; Ferrari, *My Terrible Joys,* p. 63.

23   *"Lobkowicz. Brauchitsch":* Brauchitsch, *Ohne Kampf Kein Siege,* pp. 12–13.

2. The Rainmaster

24   *"an intoxicated giant":* Nixon, *Kings of Nürburgring,* p. 15.

Only René Dreyfus: Motor, July 6, 1932; see the 1932 Eifelrennen entry at *The Golden Era of Grand Prix Racing* (website), http://www.kolumbus.fi/leif.snellman/gp3207.htm#30.

A smooth, rational, and imperturbable: Rao, *Rudolf Caracciola,* pp. 498–500.

"Rudolf Caracciola": Monkhouse, *Grand Prix Racing,* p. 43.

25   *The band at the Kakadu nightclub:* Caracciola, *A Racing Car Driver's World,* pp. 2–8; Molter, *Rudolf "Caratsch" Caracciola,* p. 27. This scene — and accompanying dialogue — is mostly drawn from Caracciola's memoir.

Just shy of six feet tall: Rao, *Rudolf Caracciola,* p. 9.

26   *an unfettered appetite:* Stuck and Burggaller, *Motoring Sport,* pp. 163–64.

The job they secured him: Molter, *Rudolf "Caratsch" Caracciola,* p. 27.

While competing: Alice Caracciola, "Memories of a Racing Driver's Wife," *Automobile Quarterly,* Summer 1968. This short memoir by Rudi's wife provides one of the finest views into the interior life of the famed driver.

After a midnight motorcycle ride: Caracciola, *A Racing Car Driver's World,* pp. 6–12, 21–26.

27 *Later that same year:* Ibid.; Molter, *Rudolf "Caratsch" Caracciola,* pp. 30–40.

"*But I want to be a racing driver*": *Road and Track,* January 1961.

Back in Dresden: Nixon, *Racing the Silver Arrows,* p. 170.

28 *Since starting at Mercedes:* Molter, *Rudolf "Caratsch" Caracciola,* pp. 30–43.

29 *Over the course of the next year:* Nixon, *Racing the Silver Arrows,* p. 170.

The concept of a self-propelled vehicle: Laux, *In First Gear,* pp. 2–4.

In the 1770s: Locomotive Engineer's Journal 25 (1991); Rolt, *Horseless Carriage,* pp. 19–22, 37.

Dozens of manufacturers across Europe: Daley, *Cars at Speed,* pp. 16–18; Ludvigsen, *Mercedes-Benz Racing Cars,* p. 8; Howe, *Motor Racing,* p. 17.

30 "*waist down*": Birkin, *Full Throttle,* p. 10.

"*Your cars are quite ugly*": Laux, *In First Gear,* p. 32.

"*This is Paris*": Howe, *Motor Racing,* p. 19.

31 *It was designed by Wilhelm Maybach:* Ludvigsen, *Mercedes-Benz Racing Cars,* pp. 14–16.

Just before dawn: Doyle, *Carlo Demand in Motion and Color,* p. 32; Helck, *Great Auto Races,* pp. 185–89; Daley, *Cars at Speed,* pp. 24–25.

"*Motor racing is unique*": Daley, *Cars at Speed,* p. 11.

Tragedy did: Autocar, July 30, 1921; Howe, *Motor Racing,* p. 22; Pomeroy, *The Grand Prix Car,* pp. 20–22.

32 *In lashing rain:* Caracciola, *A Racing Car Driver's World,* p. 35.

A month earlier: Kimes, *The Star and the Laurel,* p. 198.

33 "*with the rather pious hope*": *Car and Driver,* November 1985.

On the dark gray Sunday afternoon: Caracciola, *A Racing Car Driver's World,* pp. 35–36; *Das Auto,* July 15, 1926.

"*Quick, man*": *Road and Track,* January 1961.

34 *Rudi roared out of the curve:* Das Auto, July 15, 1926; "Mercedes beim Grossen Preis von Deutschland," 652, DBA; Caracciola, *A Racing Car Driver's World,* pp. 35–41; Neubauer, *Speed Was My Life,* pp. 5–6; Photographs, Grosser Preis von Deutschland, AVUS 1926, 93/652. DBA.

35   *"Yes," he said. "I must"*: Caracciola, *A Racing Car Driver's World*, p. 41.
      Rudi was sitting at the breakfast table: Neubauer, *Speed Was My Life*, p. 16.
      The Wall Street crash: Weitz, *Weimar Germany*, pp. 161–65.

36   *The forty-five-year-old executive*: Neue Deutsche Biographie, vol. 11, pp. 685–87.
      "modern-day Falstaff": Nixon, *Racing the Silver Arrows*, p. 188.

37   *"There was only"*: Neubauer, *Speed Was My Life*, p. 3.
      Neubauer had always been: Road and Track, May 1958; Sports Car Illustrated,
      September 1956.
      "It's finished": Neubauer, *Speed Was My Life*, p. 16.
      After Rudi and Charly: Molter, *Rudolf "Caratsch" Caracciola*, p. 37.

38   *On April 12, 1931*: Howe, *Motor Racing*, pp. 194–97; Daley, *Cars at Speed*, p.
      15.
      He had charted out every mile: Neubauer, *Speed Was My Life*, p. 22.
      In his Mercedes: Motorsport, May 1931.
      Winner the previous year: Yates, *Ferrari*, p. 53.
      "The Maestro": Collection of profile articles on Nuvolari, Personal Papers
      of Karl Ludvigsen, REVS.

39   *Dawn crested*: Neubauer, *Speed Was My Life*, pp. 26–27.
      Shortly after Rudi's return: Caracciola, *A Racing Car Driver's World*, pp.
      160–63; Neubauer, *Speed Was My Life*, pp. 29–30; Stuck and Burggaller,
      Motoring Sport, pp. 162–63; Rao, *Rudolf Caracciola*, pp. 104–10. Like many
      of the firsthand accounts that relate to Hitler, the stories change, the im-
      pressions alter, depending on if they were written before or after the war.
      Of note here, the Stuck book, published prewar, relates an article from
      Caracciola that states how impressed he was, and what an honor it was, to
      be in the presence of Hitler. His later memoir has a very different tenor.
      "All that serves": Shirer, *The Rise and Fall of the Third Reich*, p. 149.
      When officials: Mommsen, *The Rise and Fall of Weimar Democracy*, p. 315.

40   *"I've come to demonstrate it"*: Caracciola, *A Racing Car Driver's World*, p. 162.
      Musclebound, with the slitted eyes: Time, June 1, 1936; Reuss, *Hitler's Motor
      Racing Battles*, p. 57.

41   *On leaving Munich*: Stuck and Burggaller, *Motoring Sport*, p. 163. In fact,
      Caracciola wrote the meeting was "his most impressive experience."
      "I've got to drive": Caracciola, *A Racing Car Driver's World*, p. 56.
      Winning was the only thing: Caracciola, *Rennen*, pp. 67–70.
      "As light-footed": Caracciola, *A Racing Car Driver's World*, p. 57.
      They were truly: Venables, *First Among Champions*, p. 71.

3. The Speed Queen and Old Gaulish Warrior

42 *"So what if their car"*: Le Journal, January 15–21, 1932. This is an incredible series of vivid accounts by Jacques Marsillac of his journey with Lucy and Laury Schell during the Monte Carlo Rally. All quoted dialogue comes from this account.

43 *Launched in 1911:* Louche, Le Rallye Monte Carlo, p. 34.
"There comes a moment": News clip reprinted in Modern Boy's Book of Racing Cars, 1938, Personal Papers of Anthony Blight.
For the 1932 Rally: Motor, January 12, 1932; L'Auto, January 10–21, 1932.
at five in the morning: Le Journal, January 15–21, 1932.
Their black Bugatti T44: Symons, Monte Carlo Rally, pp. 5–6.

44 *A number of competitors: Motor,* January 19, 1932.

45 *The only child:* Neubauer, Speed Was My Life, p. 86.
Her father: Marriage certificate of Celestine Roudet and Francis Patrick O'Reilly, January 12, 1896, and marriage certificate of Lucy O'Reilly and Selim Schell, August 30, 1917, both in Archives de Brunoy, France.

46 *"While she grew up"*: Blight, The French Sports Car Revolution, p. 86.
"I am American": Paris Soir, May 1, 1938.
"When one is not French": Weber, The Hollow Years, p. 87.
"His life seems": Ibid.
"Even the most dangerously wounded soldiers": Reading Eagle, May 17, 1915.
Two years later: Marriage certificate of Lucy O'Reilly and Selim Schell, August 30, 1917, Archives de Brunoy, France.

47 *"Every day was like"*: Quoted in When Paris Sizzled by Mary McAuliffe in the New York Times, October 14, 2016.
The births of her children: Road and Track, February 1957; Adatto and Meredith, Delahaye Styling and Design, p. 113.
A defining characteristic: Tragatsch, Die Grossen Rennjahre 1919–1939, pp. 84–86.
Lucy followed: Bouzanquet, Fast Ladies, pp. 11–30.
In 1927, Lucy signed up for her first race: Brochure for Journée Féminine de L'Automobile, 1927, Personal Papers of Maurice Phillipe, REVS.
By the early 1930s: Blight, The French Sports Car Revolution, p. 87; April 1, 1930, Automobilia; February 25, 1929, La Vie Automobile; January 20–23, 1931, L'Auto.

Before one race: Le Journal, May 30, 1929.

"looking for trouble": Symons, *Monte Carlo Rally,* p. 4.

48   *The* pop-pop *of flashbulbs: Le Journal,* January 15–21, 1932. As before, all dialogue was sourced from the Marsillac series. In addition, the author benefited from accounts in *Motorsport,* February 1932; *Autocar,* January 12, 1932; *Motor,* January 1932; and *L'Auto,* January 10–23, 1932.

53   *"nationalist leader": Le Journal,* January 19, 1932.

The tall, distinguished man: Mays, *Split Seconds,* pp. 105–10; *Automobile Quarterly,* July 1999; *Delahaye Club Bulletin,* Winter 2014.

"Monsieur Charles": Delahaye Club Bulletin, September 1987.

"The thirties were doom-laden": Blight, *The French Sports Car Revolution,* p. 15.

Weiffenbach may not have feared: La Vie Automobile, March 10, 1931.

Bumbling politicians: Weber, *The Hollow Years,* p. 5.

The Great Depression: Malino and Wasserstein, *The Jews in Modern France,* p. 50.

54   *Madame Marguerite Desmarais: Automobile Quarterly,* Summer 1967; Bradley, *Ettore Bugatti,* p. 60; Marc-Antoine, *Delahaye 135,* pp. 11–13. The scene with Marguerite Desmarais was well documented by Pierre Peigney. The presence of Weiffenbach at the meeting was excluded. However, given the nature of the conversation and the fact that he was on the Delahaye board at the time (and its long-running leader), it is questionable that the family would have conducted this meeting without his input, particularly since he was leading the charge for mass production.

"Midrange touring cars": Delahaye Club Bulletin, Winter 2014.

Unlike other early French manufacturers: Delahaye Club Bulletin, January 2016.

The two-cylinder engine: Beadle, *Delahaye,* p. 5; *Motorsport,* October 1936.

55   *At its debut:* Mays, *Split Seconds,* pp. 13–18; Laux, *In First Gear,* p. 59; *La Locomotion Automobile,* vol. 2, 1895.

The 1896 race: Motorsport, October 1936; Helck, *Great Auto Races,* p. 11.

Late the following year: Mays, *Split Seconds,* pp. 18–23.

56   *The company blossomed:* The 3.5 Litre Delahaye Type 135 (profile publications), PPRA.

"Solid as a Delahaye": Torque, January–February 1984.

Following his instinct: Jolly, *Delahaye: Sport et Prestige,* pp. 8–15.

He was a no-nonsense leader: L'Auto, October 2, 1936.

"The typical product": Draft of "The 1936 Delahaye Type 135 Competition," Personal Papers of Karl Ludvigsen, REVS.

"provincial notary": Jolly, *Delahaye: Sport et Prestige,* p. 6.

57 *"a Benzedrine tablet":* Beadle, *Delahaye: Road Test Portfolio,* p. 72.

"Charles," he said: Adatto and Meredith, *Delahaye Styling,* p. 213; Dorizon, Peigney, and Dauliac, *Delahaye,* p. 33.

"from the shadows": Blight, *The French Sports Car Revolution,* p. 57.

The sun was shining brightly at Montlhéry: Ibid., p. 80.

58 *The strange vehicle: Automobile,* March 1985.

The "son of a truck" engine: Torque, January 1984; *Omnia,* April 1935.

Wearing a black cloth helmet: Mays, *Split Seconds,* pp. 107–13; Marc-Antoine, *Delahaye 135,* pp. 14–16.

"What on earth": Blight, *The French Sports Car Revolution,* p. 80.

Held in the Grand Palais: September 26, 1933–October 14, 1933, *L'Auto;* Draft of "The 1936 Delahaye Type 135 Competition," Personal Papers of Karl Ludvigsen, REVS; *Le Fanatique de L'Automobile,* April 1976.

Weiffenbach introduced: La Vie Automobile, November 1933.

59 *The first version:* The actual names of these first models are somewhat ambiguous. Most historians label them Types 134 and 138, but Delahaye only referred to them at the Salon as 12CV and 18CV, respectively, a reference to their taxable government rating for horsepower, a convoluted formula that would be meaningless to describe. For the sake of clarity, I refer to them as 134 and 138 and have altered the dialogue from Lucy Schell as featured in Blight's book accordingly, for consistency's sake.

"Their names?": Blight, *The French Sports Car Revolution,* pp. 84–85.

60 *Weiffenbach neglected:* October 31, 1933, *L'Auto.*

4. Crash

63 *René Dreyfus seemed: La Dépêche,* August 15, 1932; *Motorsport,* September 1932; *L'Auto,* August 15, 1932; *L'Automobile sur la Côte d'Azur,* January 1934.

64 *"Did you win?":* Original interviews between René Dreyfus and Beverly Kimes, PPBK.

"Thank you very much": Dreyfus and Kimes, *My Two Lives,* p. 34.

"hovered around us": Michel Ribet, interview with the author, Comminges, France, 2018.

65 *"bone collectors"*: Caracciola, *Rennen,* p. 53.

They had seen it so often: Undated news clip, René Dreyfus Scrapbooks, MMA.

They measured how fast: Daley, *Cars at Speed,* p. 187.

They could not perform: Unsourced and undated interview with René Dreyfus, René Dreyfus Scrapbooks, MMA.

"to show the wood": Brauchitsch, *Ohne Kampf Kein Siege,* p. 46.

René loved the life: *Sports Car Graphic,* September 1984; *Road and Track,* September 1980; Ribet, interview with the author.

The community of drivers: *L'Intransigent,* July 18, 1933; Ruesch, *The Racer,* p. 54; Court, *A History of Grand Prix Motor Racing,* p. 111; Stuck, *Männer hinter Motoren,* pp. 36–38.

Among this band: *Motor,* June 22, 1937; Ribet, interview with the author; Kimes, *The Star and the Laurel,* p. 220; Dreyfus and Kimes, *My Two Lives,* p. 15; Neubauer, *Speed Was My Life,* pp. 20–21; Birabongse, *Bits and Pieces,* pp. 36–37.

66 They shared a strange kind of friendship: Ferrari, *My Terrible Joys,* p. 89.

"Let's not be sentimental": Neubauer, *Speed Was My Life,* p. 18.

Above everything: *L'Automobile sur la Côte d'Azur,* January 1934.

"I'm ready": Original interviews between Dreyfus and Kimes, PPBK.

Sporting his trademark brown bowler hat: Bugatti, *The Bugatti Story,* p. 16.

67 Early in 1933: *Automobile Quarterly,* Summer 1967; Dreyfus and Kimes, *My Two Lives,* p. 37.

The restaurants, grocery stores: Bugatti, *The Bugatti Story,* pp. 29–30.

It was this attention: Dreyfus and Kimes, *My Two Lives,* p. 41.

68 On February 11: Domarus, *Hitler,* p. 250.

"In massive columns": François-Poncet, *The Fateful Years,* p. 48.

The new chancellor: Bullock, *Hitler,* pp. 110–13.

He called for new Reichstag elections: Shirer, *The Rise and Fall of the Third Reich,* p. 190.

"These momentous tasks": Domarus, *Hitler,* p. 251.

69 The purpose of the NSKK: Hochstetter, *Motorisierung und "Volksgemeinshaft,"* pp. 481–82.

"defensive force of the nation": Ibid., p. 242.

70 *"force, strength"*: Reuss, *Hitler's Motor Racing Battles,* p. 51.

"personal confidant": Pohl, Habeth-Allhorn, and Brüninghaus, *Die Daimler-Benz AG in den Jahren,* pp. 36–38.

Board members: Gregor, *Daimler-Benz in the Third Reich,* pp. 57–58.

"no occasion": Bellon, *Mercedes in Peace and War,* p. 219.

With car and truck sales: Ibid., pp. 216–17.

orders for heavy trucks: Minutes of the Daimler-Benz board meeting, April 25, 1933, Kissel Files, DBA.

If they moved toward: Gregor, *Daimler-Benz in the Third Reich,* pp. 58–61.

71 *"You will receive":* Brauchitsch, *Ohne Kampf Kein Siege,* p. 28.

"Reich racing cars": Reuss, *Hitler's Motor Racing Battles,* p. 43.

Auto Union: Pohl, Habeth-Allhorn, and Brüninghaus, *Die Daimler-Benz AG in den Jahren,* p. 108.

"profound speech": Reuss, *Hitler's Motor Racing Battles,* p. 67.

At the March 10 board meeting: Minutes of March 10, 1933, Daimler-Benz board meeting, Kissel Files, DBA.

"Highly Esteemed Herr Reich Chancellor": Reuss, *Hitler's Motor Racing Battles,* p. 69.

Although Daimler-Benz: Minutes of March 10, 1933, Daimler-Benz board meeting, Kissel Files, DBA.

72 *On Thursday, April 20, 1933:* L'Auto, April 21, 1933; *L'Intransigent,* April 22, 1933.

American-born: Rao, *Rudolf Caracciola,* pp. 297–98.

"French charm, German perseverance": Neubauer, *Speed Was My Life,* p. 97.

73 *"Rudi's silent love":* Automobile Quarterly, Summer 1968.

"You know": Caracciola, *A Racing Car Driver's World,* p. 60.

Thus, "Scuderia CC": Nixon, *Racing the Silver Arrows,* p. 75.

By their last run: Caracciola, *Rennen,* pp. 74–75.

He cranked down: Motorsport, June 1933; *Autocar,* April 28, 1933; *L'Auto,* April 21, 1933.

74 *Rudi was carried:* Caracciola, *A Racing Car Driver's World,* p. 62.

75 *"Tell them to pull my leg":* Ibid.

"Look, madame": Ibid., p. 64.

In the final lap: Autocar, April 28, 1933.

Despite everything: Caracciola, *A Racing Car Driver's World,* pp. 63–64.

76 *"Do not ask him":* L'Intransigent, April 25, 1933.

After his third-place finish: Automobile Quarterly, Summer 1967; Dreyfus race record, PPBK; Belle, *Blue Blood,* p. 86; Saward, *The Grand Prix Saboteurs,* p. 66.

Nicknamed "Chou-Chou": Pierre Fassone, telephone interview with Jonathan Dupriez, France, 2018; Dreyfus-Miraton marriage record, Archives de Brunoy.

Chou-Chou and René crossed paths: Evelyne Dreyfus, interview with the author.

One evening: Dreyfus and Kimes, *My Two Lives,* p. 27.

77 *The conductor Arturo Toscanini:* Sachs, *The Letters of Arturo Toscanini,* p. 127.

René had lived: Dreyfus and Kimes, *My Two Lives,* p. 27.

"anti-Semitism of principle": Weber, *The Hollow Years,* pp. 102–3.

"One did not have to think ill": Ibid., p. 103.

Despite this underlying anti-Semitism: Evelyne Dreyfus, interview with the author; Dreyfus and Kimes, *My Two Lives,* p. 64. René distinctly declares in his autobiography that he was distantly related to Captain Dreyfus. By contrast, subsequent research by later family members have determined otherwise.

"They know there is a menace": Weber, *The Hollow Years,* p. 243.

78 *On September 10: Automobile Quarterly,* Summer 1967; Dreyfus and Kimes, *My Two Lives,* p. 4.

Giuseppe Campari: Motor, September 12, 1933; Seymour, *Bugatti Queen,* p. 175; see the 1933 Monza entry at *The Golden Era of Grand Prix Racing* (website), http://www.kolumbus.fi/leif.snellman/gp3314.htm#64.

The Italian leader: Yates, *Ferrari,* p. 36.

79 *"Start—and win!" he ordered his drivers:* Cernuschi, *Nuvolari,* p. 61.

Now three more lives: L'Auto, September 12–17, 1933; *Motor,* September 19, 1933; Louche, *1895–1995,* pp. 207–8.

René had been closest: L'Auto, September 12–17, 1933.

Also listening to the race that day: Caracciola, *A Racing Car Driver's World,* pp. 66–67.

Midseason, Louis Chiron: Cernuschi, *Nuvolari,* pp. 83–93; *Motorsport,* December 1933.

"Well, it'll be all right": Caracciola, *A Racing Car Driver's World,* pp. 71–75.

80 *"Of course, I can":* Ibid.; Caracciola, *Rennen,* pp. 75–76; Neubauer, *Speed Was My Life,* pp. 52–53. As was typical, Neubauer tended to conflate scenes. Taken with Caracciola's two memoirs, the author has done his best to separate the conversations and meeting that occurred between November 1933 and January 1934.

81 *"Fit and well again?":* Caracciola, *A Racing Car Driver's World,* pp. 74–75; Neubauer, *Speed Was My Life,* pp. 52–53.

The next day: Caracciola, *Rennen,* pp. 75–76.

Dr. Hans Nibel: Motor, February 1, 1938; Ludvigsen research notes, Personal Papers of Karl Ludvigsen, REVS; Walkerly, *Grand Prix,* p. 20; Stuck and Burggaller, *Motoring Sport,* pp. 10–11.

82 *"cars, even the somewhat advanced P3":* Yates, *Ferrari,* p. 84.

The prototype Mercedes engine: "Der Mercedes-Benz Rennwagen, ein Meisterwerk deutscher Technik" (internal Mercedes-Benz report), DBA; Ludvigsen research notes, Personal Papers of Karl Ludvigsen, REVS; *Road and Track,* December 1971. For those seeking further information on the development of the W25, this *Road and Track* article by Karl Ludvigsen, the preeminent Mercedes-Benz historian, is a minutely detailed investigation of its design and characteristics.

On February 2: Automobile Quarterly, Summer 1968; Caracciola, *A Racing Car Driver's World,* pp. 76–77; *L'Auto,* February 5, 1934.

5. The One Thing

84 *Past them strode:* Blight, *The French Sports Car Revolution,* p. 91.

Perrot had placed: L'Auto, January 28, 1934.

Lucy did not: Paris–Saint-Raphaël Rally, winners list, Personal Papers of Maurice Phillipe, REVS; Blight, *The French Sports Car Revolution,* p. 91. According to Blight, "The two Works Delahayes [driven by Nenot and Gonnot?] finished first and second in the class for engines over 17CV, but [being] penalized by the complicated handicap system under which the Rally was run, finished well below Lucy in the final order."

While Lucy fought her battle: Weber, *The Hollow Years,* pp. 114–34.

85 *"It appears that France":* Stobbs, *Les Grandes Routieres,* p. 17.

While women in America: Weber, *The Hollow Years,* pp. 76–83.

In motorsport, Lucy was up against: Bullock, *Fast Women,* pp. xi–xii.

"feminine excitability": Bouzanquet, *Fast Ladies,* preface.

"weak and delicate": La Vie Automobile, September 25, 1929.

86 *"only attentive to the aesthetic factor":* news clip of undated editorial, René Dreyfus Scrapbooks, MMA.

"They chase after us": Ribet, interview with the author.

"best Delahaye rally driver": Blight, *The French Sports Car Revolution,* p. 91.

There was another factor: Dreyfus and Kimes, *My Two Lives,* p. 70.

On Saturday, March 24: Automobile sur la Côte d'Azur, January/March 1934; *L'Auto,* March 21–29, 1934.

87 *"his head vanished":* Griffith Borgeson, "The Zborowski Saga," *Automobile Quarterly,* 2nd quarter 1984.

"the demeanor of a barroom fighter": Road and Track, August 1984.

"admirable virtuosity": L'Intransigent, April 1, 1934.

88 *The two barely knew each other:* Dreyfus and Kimes, *My Two Lives,* p. 69.

The young engineer: Marc-Antoine, *Delahaye 135,* pp. 20–22; Tissot, *Figoni Delahaye,* pp. 13–16.

Montlhéry was a favorite site: Blight, *The French Sports Car Revolution,* p. 91; Nixon, *Racing the Silver Arrows,* p. 56.

89 *At 4:00 p.m.:* L'Auto, May 8–11, 1934.

At lunchtime on May 10: Blight, *The French Sports Car Revolution,* pp. 92–93.

90 *A weary, hood-eyed François:* L'Auto, May 11, 1934.

Raised in Revel: Mays, *Split Seconds,* p. 109; *Automobile Quarterly,* July 1999.

"old French firm": L'Auto, May 11, 1934.

"They used to say": L'Intransigent, May 16, 1935.

After Charly's funeral: Automobile Quarterly, Summer 1968.

91 *In early April:* Rao, *Rudolf Caracciola,* p. 163.

When he returned to the harbor: see the 1934 Monaco entry at *The Golden Era of Grand Prix Racing* (website), http://www.kolumbus.fi/leif.snellman/gp3401.htm#3.

"For me, there had to be": Caracciola, *A Racing Car Driver's World,* p. 80.

While Rudi prepared: Nixon, *Racing the Silver Arrows,* p. 26; Neubauer, *Speed Was My Life,* p. 55.

He asked for the dawn start: Rao, *Rudolf Caracciola,* pp. 173–74.

Now sheathed in a sleek white-painted aluminum body: Road and Track, December 1971.

He looked every inch: Rao, *Rudolf Caracciola,* pp. 174–75.

92 *One Mercedes driver likened it:* Walkerly, *Grand Prix,* p. 29.

"You did 235 [kph]": Caracciola, *A Racing Car Driver's World,* p. 82.

93 *"decrepit old cars":* Motorsport, August 1934.

"strange silhouette": L'Auto, November 21, 1933.

The P-Wagen (P[orsche]-car): L'Auto, March 7, 1934; Cancellieri, *Auto Union,* p. 39.

"the tearing exhaust note": Motorsport, April 1934.

94   *"his country's cars to be supreme"*: Dreyfus and Kimes, *My Two Lives,*
p. 52.

As for his design: Road and Track, August 1976.

"Well, it's too bad": *Automobile Quarterly,* Summer 1967.

"scientifically designed to include": *Motor Trend,* March 1959.

With their drivers seated: Autocar, July 6, 1934; *Motorsport,* August 1934.

"We shall win tomorrow": *Motor Trend,* March 1959.

95   *"Bloody Day of Repression"*: *Le Matin,* July 1, 1934.

According to wire reports: Ibid.; *L'Intransigent,* July 1, 1934; *New York Times,*
July 1, 1934; Bullock, *Hitler,* pp. 122–30.

"alarming news": Brauchitsch, *Ohne Kampf Kein Siege,* p. 29.

"the noisiest car on earth": Ludvigsen, *Mercedes-Benz Racing Cars,* p. 121.

After the starting flag fell: Lyndon, *Grand Prix,* pp. 168–81; *L'Auto,* July 3,
1934; *Motor,* July 3, 1934; *Autocar,* July 6, 1934.

96   *"The mighty German assault"*: *Motor Trend,* March 1939.

under an overcast July sky: Motorsport, August 1934; Cernuschi, *Nuvolari,* p.
122; *Motor,* March 16, 1937; *Autocar,* August 2, 1935.

97   *"never bending, never capitulating"*: Feldpost No. 42, February 7, 1942, NS
24/846, Bundesarchiv; Hamilton, *Leaders and Personalities of the Third Reich,*
pp. 287–88; Hilton, *How Hitler Hijacked World Sport,* p. 12; Hochstetter,
Motorisierung und "Volksgemeinshaft," p. 124.

"man of action": Reuss, *Hitler's Motor Racing Battles,* p. 101.

"battle for the motorization of Germany": Hochstetter, *Motorisierung und
"Volksgemeinshaft,"* p. 2.

98   *"the cavalry of the future"*: Ibid., p. 101.

In return, the Nazi government: "Bericht uber Kosten fur den Bau und die
Entiwckelung eines neuen Rennwagentyps," November 9, 1934, DBA;
Reuss, *Hitler's Motor Racing Battles,* pp. 81–83. In his well-researched book,
Reuss details how Mercedes and Auto Union received far more than previ-
ously reported. The Mercedes-Benz archive records match his findings.

The symbiotic relationship: Shirer, *The Rise and Fall of the Third Reich,* pp.
281–83; Gregor, *Daimler-Benz in the Third Reich,* pp. 36–75.

After the blast of a military cannon: Caracciola, *A Racing Car Driver's World,* p.
83.

"motor racing is no longer": Hochstetter, *Motorisierung und "Volksgemeinshaft,"*
p. 292.

His W25 was running faster: Road and Track, December 1971.

99 *Through the first dozen laps: Motorsport,* August 1934.

During the Italian Grand Prix: Motorsport, October 1934; Caracciola, *A Racing Car Driver's World,* pp. 81–92; Neubauer, *Speed Was My Life,* p. 58.

100 *As the 1934 season approached an end:* Original interviews between Dreyfus and Kimes, PPBK; *Motorsport,* September 1934.

He loved being part of Bugatti: Road and Track, August 1976.

"the Old Man": René Dreyfus, letter to Maurice Louche, 1990, PPML.

Journalist Georges Fraichard: Match, April 3, 1934.

When René was honest with himself: Dreyfus and Kimes, *My Two Lives,* p. 51.

René was so fed up: Ibid., p. 50; undated Maurice Louche interview with René Dreyfus, PPML.

101 *"We would like": L'Auto,* August 26, 1934.

Every inch of the cobblestoned city: Cholmondeley-Tapper, *Amateur Racing Driver,* pp. 47–48; Chakrabongse, *Road Start Hat Trick,* p. 137.

Although everyone: Dreyfus and Kimes, *My Two Lives,* p. 52.

On August 26: Autocar, August 31, 1934; *L'Auto,* August 27, 1934.

"No, Dreyfus was second": Dreyfus and Kimes, *My Two Lives,* p. 53.

To quiet the violence: Original interviews between Dreyfus and Kimes, PPBK; *Automobile Quarterly,* Summer 1968.

A few days later: Bradley, *Ettore Bugatti,* p. 60.

"René, you could be one of the greatest drivers in the world": Sports Car Guide, September 1959; Dreyfus and Kimes, *My Two Lives,* p. 51; *Automobile Quarterly,* Summer 1967. Dreyfus recounts this same conversation almost verbatim in these three separate publications. The author has folded the thrust of Costantini's advice into this single statement.

102 *René departed Molsheim:* Original interviews between Dreyfus and Kimes, PPBK.

Two weeks before their December 8 wedding: Certificat de mariage, Dreyfus et Miraton, Archives de Brunoy.

Chou-Chou and her father: Dreyfus and Kimes, *My Two Lives,* pp. 53–54.

Prenuptial agreements: Ibid., p. 54.

6. The Shadow

104 *In May 1935:* Caracciola, *A Racing Car Driver's World,* pp. 107–19.

"Racing is and always will be": Hochstetter, *Motorisierung und "Volksgemeinschaft,"* p. 308.

105 *"the benchmark for the industrial ability"*: Ibid.

"a community of prodigious men": Day, *Silberpfeil und Hakenkreuz*, p. 95.

In March 1935: Shirer, *The Rise and Fall of the Third Reich*, p. 284.

"Weren't they all communists": Brauchitsch, *Ohne Kampf Kein Siege*, pp. 101–4. These are some remarkably disturbing passages from Brauchitsch, but they provide an excellent window into the mind-set of the drivers in how they thought about the crimes of the state.

106 *The star-shaped Uaddan:* Lang, *Grand Prix Driver*, p. 36.

"fourth shore": Moretti, *Grand Prix Tripoli*, p. 10.

"political Fascist thug": Ibid.

The course's huge cantilevered grandstand: Daley, *Cars at Speed*, p. 148.

"a fine sense of balance and touch": Monkhouse, *Grand Prix Racing*, p. 42.

"fastest road circuit in the world": *Motorsport*, June 1935.

Rudi suffered three tire failures: Cernuschi, *Nuvolari*, p. 116; Caracciola, *A Racing Car Driver's World*, pp. 107–18.

"Kaiser Wilhelm's army of 1914": *Speed*, August 1937.

When a Mercedes driver needed: Lang, *Grand Prix Driver*, pp. 41–42.

107 *"No. 1 gets the left rear wheel ready"*: Neubauer, *Speed Was My Life*, p. 156.

"the secret of victory": Caracciola, *Rennen*, p. 83.

With five laps remaining: Ibid.; *Motor*, May 21, 1935; *L'Auto*, May 12, 1935; see the 1935 Tripoli entry at *The Golden Era of Grand Prix Racing* (website), http://www.kolumbus.fi/leif.snellman/gp3502.htm#6.

108 *"There was the sun"*: Caracciola, *A Racing Car Driver's World*, p. 118.

"absolute superiority": "Mercedes-Benz Press Informationsdienst, Tripoli 1935," MB 128/1031, DBA.

A month later, in mid-June: Neubauer, *Speed Was My Life*, pp. 70–71; *Motorsport*, July 1935; see the 1935 Eifelrennen entry at *The Golden Era of Grand Prix Racing* (website), http://www.kolumbus.fi/leif.snellman/gp3504.htm#19.

As was expected of him: Day, *Silberpfeil und Hakenkreuz*, p. 85.

109 *"Well done, my dear boy"*: Rao, *Rudolf Caracciola*, p. 210.

He was more vainglorious: Hilton, *Hitler's Grands Prix in England*, p. 74.

At the time, one journalist: Ibid.

Throughout the summer of 1935: *Road and Track*, October 1983; Dreyfus and Kimes, *My Two Lives*, p. 55.

After the race: See the 1935 Monaco entry at *The Golden Era of Grand Prix Racing* (website), http://www.kolumbus.fi/leif.snellman/gp3501.htm#3.

Afterward, a newspaper cartoon: Undated newspaper cartoon, French Grand Prix 1935, Race files (1037), DBA.

"the great misery": La Vie Automobile, July 10, 1935.

110 *"When will it be understood":* L'Intransigent, June 25, 1935.

The ACF took the coward's path: Paris and Mearns, *Jean-Pierre Wimille,* p. 67.

At the Belgian Grand Prix: Motorsport, August 1935.

René spent so much time: L'Auto, August 1, 1935. The effect of the WW fuel prompted a big debate, and Mercedes-Benz even called a press conference to refute the claims.

111 *Thirty-seven years old:* Yates, *Ferrari,* pp. x–xi.

While Nuvolari was all emotion: Ferrari, p. 63.

The two could not stand: "1935 Season Lineup," *The Golden Era of Grand Prix Racing* (website), http://www.kolumbus.fi/leif.snellman/gp3501.htm#SL.

112 *"You sir":* Cernuschi, *Corse per Il Mondo,* p. 297.

Away from races: Road and Track, September 1980.

Five feet, five inches tall: Motor, June 22, 1937.

"He drove like a madman": Purdy, *The Kings of the Road,* pp. 43–49.

Wearing his trademark lemon-yellow, sleeveless jersey: Cholmondeley-Tapper, *Amateur Racing Driver,* p. 93.

113 *"Corri! [Come on!]":* Carter, *Nuvolari and Alfa Romeo,* n.p.

After the first lap: Cernuschi, *Nuvolari,* p. 123.

Round and round: L'Auto, August 1, 1935; *Autocar,* August 2, 1935; *Motor,* July 30, 1935; Canestrini, *Uomini E Motori,* pp. 188–91; *Motorsport,* September 1935; see the 1935 German Grand Prix entry at *The Golden Era of Grand Prix Racing* (website), http://www.kolumbus.fi/leif.snellman/gp3507.htm#31.

"inspired, fearless, untouchable": Cernuschi, *Nuvolari,* p. 126.

"remorseless dance of death": Ibid., p. 128.

114 *"Brauchitsch is being closely followed":* Hilton, *Nuvolari,* p. 154.

"bad luck": Herzog, *Unter dem Mercedes-Stern,* pp. 52–53.

"once again that man": Cernuschi, *Nuvolari,* p. 130.

Situated eighteen miles southeast of Paris: Gautier and Altounian, *Memoire en Images Brunoy,* pp. 7–126.

115 *But the two-seater convertible roadster:* Mays, *Split Seconds,* p. 242. For the sake of clarity, the author has decided to refer to the Delahaye 135 Special (the

sports-car version produced for Lucy) as simply "the 135," since this is the only version profiled in the book.

At 3.5 liters: Marc-Antoine, *Delahaye 135,* pp. 26–28; Blight, *The French Sports Car Revolution,* p. 147.

Before François finished building the 135: L'*Auto,* December 17, 1935; L'*Auto,* October 9, 1935.

"There you are": Blight, *The French Sports Car Revolution,* p. 151.

Every detail: La Rairie, Credit Agricoles Papers, Archives de Brunoy.

As well as preparing: Blight, *The French Sports Car Revolution,* pp. 150–51.

116   *"Put Nuvolari in the car":* Zagari, *Tazio Nuvolari,* p. 9.

His countrymen: Motorsport, October 1935; see the 1935 Italian Grand Prix entry at *The Golden Era of Grand Prix Racing* (website), http://www.kolum bus.fi/leif.snellman/gp3509.htm#39.

The Italian crowd: L'*Auto,* September 9–10, 1935.

Worse were the sneers: Ribet, interview with the author.

Signs posted outside restaurants: Wolff, *The Shrinking Circle,* p. 12.

At Nuremberg: Burden, *The Nuremberg Rallies,* pp. 108–10.

117   *"Politics are normally": Motorsport,* December 1935.

With several wins: See "AIACR European Championship 1935," *The Golden Era of Grand Prix Racing* (website), http://www.kolumbus.fi/leif.snellman/ cha5.htm; *Road and Track,* October 1983; *Road and Track,* August 1976.

"continual agitation": Motorsport, May 1935.

The annual shuffle of drivers: Automobile Quarterly, 2nd quarter 1979.

118   *"Your name is Dreyfus":* Dreyfus and Kimes, *My Two Lives,* p. 120. Although said by a Vichy judge during the war, the sentiment was no doubt the same expressed earlier by such racist ideologues.

apart from Nuvolari: Ribet, interview with the author; *Motorsport,* January 1936; Stevenson, *Driving Forces,* pp. 138–50; Tragatsch, *Die Grossen Rennjahre 1919–1939,* pp. 259–60; Yates, *Ferrari,* pp. 108–9.

Adolf Rosenberger: Don Sherman, "Porsche's Silent Partner," *Hagerty,* August 9, 2018, https://www.hagerty.com/articles-videos/arti cles/2018/08/09/the-story-of-adolf-rosenberger.

"They will decide": Dreyfus and Kimes, *My Two Lives,* p. 87.

119   *"splendid": ACF Bulletin Officiel,* October 1935.

With a jubilant shout: La Vie Automobile, February 25, 1936; *Delahaye Club Bulletin,* March 2002; Symons, *Monte Carlo Rally,* pp. 208–9.

The warm weather: Horizons, March 1936; *L'Automobile sur la Côte d'Azur,* February 1936.

120 *Their Delahaye: Motorsport,* August 1936.

"The whole machine": Autocar, September 1936.

When the Schells: Horizons, March 1936; *L'Automobile sur la Côte d'Azur,* February 1936.

That year: La Vie Automobile, February 25, 1936; *Motorsport,* February 1936.

121 *The next month: L'Auto,* March 7–12, 1936.

Lucy suffered a further poor performance: L'Auto, April 1936; *La Vie Automobile,* May 10, 1936; Blight, *The French Sports Car Revolution,* pp. 177–83.

On February 15: Motorsport, March 1936; Nixon, *Racing the Silver Arrows,* p. 108.

122 *"no longer be automobiles": L'Auto,* February 15, 1936.

After the debacle: September 1937, Archives de Brunoy; *La Vie Automobile,* November 10, 1934; *La Vie Automobile,* May 10, 1935; *L'Auto,* December 17, 1935; Blight, *The French Sports Car Revolution,* p. 153.

123 *"France wants peace":* Weber, *The Hollow Years,* pp. 145–46.

"Hitler has got away with it!": Shirer, *Berlin Diary,* pp. 55–56.

In the wake: René Dreyfus, interview with Jean Paul Caron, 1973, PPRA; *Paris Soir,* May 1, 1938.

124 *"What now, Madame Schell?":* Blight, *The French Sports Car Revolution,* p. 175; Dreyfus, interview with Caron, 1973.

7. A Very Good Story

125 *As a young man in his twenties:* Mays, *Split Seconds,* p. 24; Legion of Honor records, Paris, France.

126 *"the marque better known": Delahaye Club Bulletin,* Winter 2014.

Most of all: Le Figaro, October 13, 1937; Dreyfus and Kimes, *My Two Lives,* p. 70.

They both knew: L'Auto, February 15, 1936.

Weiffenbach added that: Dorizon, Peigney, and Dauliac, *Delahaye,* pp. 50–52.

Over the next few days: Venables, *French Racing Blue,* pp. 128–29; Strother MacMinn, "Delahaye Type 145 Coupe," PPRA.

In the spring of 1936: Abeillon, *Talbot-Lago de Course,* p. 7.

127 *"We still have a lot to do": L'Intransigent,* December 18, 1935.

Lago was a garrulous, handsome Italian: Larsen, *Talbot-Lago Grand Sport,* pp. 17–20.

Formed in 1903: Ibid., pp. 19–21; *Automobile Quarterly,* Spring/Summer 1965; *Automobile Quarterly,* 4th quarter 1985.

128 *"in the very near future":* L'Auto, February 7, 1935.

This would include: Blight, *The French Sports Car Revolution,* p. 149.

On May 24: Automobile Quarterly, 4th quarter 1985.

"Just go as fast as you can": Dreyfus and Kimes, *My Two Lives,* p. 64.

René followed the orders: L'Auto, May 25, 1936.

129 *Two days after the race:* Blight, *The French Sports Car Revolution,* pp. 192–93; Weber, *The Hollow Years,* pp. 150–53.

No corner of France: Larsen, *Talbot-Lago Grand Sport,* pp. 20–24.

"I've nothing to worry about": Bugatti, *The Bugatti Story,* p. 82.

Wider and longer: Blight, *The French Sports Car Revolution,* pp. 197–201.

"Your job will be": Automobile Quarterly, Spring/Summer 1965.

At ten o'clock: Motorsport, August 1936; *L'Auto,* June 25–30, 1936.

130 *At the Marne race:* L'Auto, July 6, 1936.

"Why, yes, I believe so": Blight, *The French Sports Car Revolution,* p. 223.

Ever since they shared: Dreyfus, interview with Louche, PPML.

"Women": Original interviews between Dreyfus and Kimes, PPBK; *Automobile Quarterly,* 4th quarter 1978.

131 *The son of a famous and well-heeled:* Paris and Mearns, *Jean-Pierre Wimille,* pp. 13–30.

On August 9: Dreyfus and Kimes, *My Two Lives,* pp. 67–68.

On May 10, 1936: Motorsport, June 1936; see the 1936 Tripoli Grand Prix entry at *The Golden Era of Grand Prix Racing* (website), http://www.kolum bus.fi/leif.snellman/gp3603.htm#8.

Watching with Governor-General Italo Balbo: Reuss, *Hitler's Motor Racing Battles,* p. 262.

132 *"bloody swine":* Neubauer, *Speed Was My Life,* pp. 76–77; Reuss, *Hitler's Motor Racing Battles,* pp. 262–68. In his history, Reuss provides fairly clear documentation that the race was a setup from the start.

She was always full of energy: Caracciola, *A Racing Car Driver's World,* p. 120.

"If I could find a girl like you": Car and Driver, November 1985.

"chalk-white Moorish façades": Neubauer, *Speed Was My Life,* pp. 77–78.

133 *"A toast to the victor":* Reuss, *Hitler's Motor Racing Battles,* p. 266.

He remained ambiguous: Brauchitsch, *Ohne Kampf Kein Siege,* pp. 118–19.

"swift as greyhounds": Day, *Silberpfeil und Hakenkreuz,* p. 137.

A blanket of fog settled: Nixon, *Kings of Nürburgring,* p. 88; see the 1936 Eifel-rennen entry at *The Golden Era of Grand Prix Racing* (website), http://www.kolumbus.fi/leif.snellman/gp3605.htm#15.

134   *"It must be a drive":* Autocar, July 1936.

"thunderbolt known as Rosemeyer": Motor, June 1936.

Bernd had been watching: Motorsport, September 1935.

135   *"Just as a bird":* Rosemeyer and Nixon, *Rosemeyer!,* p. 27.

"Like a high-speed camel": Ibid., p. 36.

"the fastest couple": Cancellieri, *Auto Union,* p. 83.

"If [he] had not existed": Pritchard, *Silver Arrows in Camera,* p. 161.

While he was racing motorcycles: Frilling, *Elly Beinhorn und Bernd Rosemeyer,* pp. 40–41; Hilton, *How Hitler Hijacked World Sport,* p. 42.

"the radiant boy": Day, *Silberpfeil und Hakenkreuz,* pp. 172–75.

136   *"Beautiful blond Bernd":* Ibid.

"Unforgettable, dazzling Bernd": Ibid.

"German heroine": Ibid., pp. 175–83; Frilling, *Elly Beinhorn und Bernd Rose-meyer,* p. 67.

Thirteen days after Bernd and Elly's marriage: Autocar, July 31, 1936.

On August 1: L'Auto, August 2, 1936; François-Poncet, *The Fateful Years,* pp. 203–6; Kershaw, *Hitler,* p. 6.

137   *"Champagne flowed like water":* Keys, *Globalizing Sport,* p. 152.

"Face contorted": François-Poncet, *The Fateful Years,* p. 205.

"truly difficult to endure": Keys, *Globalizing Sport,* p. 153.

JEWS AND ANIMALS NOT ALLOWED: Ibid., pp. 134–42.

"witnessing the revival": Klemperer, *I Will Bear Witness,* p. 180.

138   *"warriors for Germany":* Keys, *Globalizing Sport,* p. 129.

"great lesson": L'Auto, August 1, 1936.

"A car is not a thing": Setright, *The Designers,* p. 13.

139   *"any more cylinders":* Ludvigsen, *The V12 Engine,* p. 13.

Although complex: Classic and Sportscar, August 1992.

"a peculiar pulse": Ludvigsen, *The V12 Engine,* p. 9.

Using the proven method: Le Fanatique de L'Automobile, April 1978; Dorizon, Peigney, and Dauliac, *Delahaye,* pp. 56–58; Strother MacMinn, "Delahaye Type 145," PPRA; Mays, *Split Seconds,* pp. 264–66; Ludvigsen, *The V12 Engine,* pp. 211–12.

regardless of how well built and tuned: Jolly, *Delahaye: Sport et Prestige*, p. 130.

140   *In its design:* Ibid.; *Le Fanatique de L'Automobile*, April 1978.

"No wild innovations here": Strother MacMinn, "Delahaye Type 145," PPRA.

That September: Autocar, September 11, 1936.

Since spinning out of control: Blight, *The French Sports Car Revolution*, p. 183.

141   *After returning to France:* Ibid., p. 271.

"finesse and intelligence": L'Auto, July 16, 1933, and May 10, 1936.

"calm, measured of movement": Moteurs Courses, 3rd quarter 1956.

Lucy liked to conduct her business: Auto Retro, September 1981.

142   *"It is to be professional":* Dreyfus and Kimes, *My Two Lives*, p. 69; Dreyfus, interview with Caron, 1973.

143   *"who talked a very good story":* Dreyfus and Kimes, *My Two Lives*, p. 70.

8. Rally

147   *"Sheep in wolves' clothing": Miniature Auto*, August 1966.

"A Jewish slave": Hochstetter, *Motorisierung und "Volksgemeinshaft,"* p. 295; *Car and Driver*, November 1985.

In the run-up: Motor, September 22, 1936.

In the first two laps: L'Auto, August 24, 1936; Neubauer, *Speed Was My Life*, p. 75; see the 1936 entry for the Swiss Grand Prix at *The Golden Era of Grand Prix Racing* (website), http://www.kolumbus.fi/leif.snellman/gp3609.htm#30.

148   *"Well, young man":* Rosemeyer and Nixon, *Rosemeyer!*, p. 85.

149   *"a remarkable driver": L'Auto*, September 15, 1936.

He wanted Baby: Automobile Quarterly, Summer 1968; Caracciola, *A Racing Car Driver's World*, pp. 119–20.

He read how: Rosemeyer and Nixon, *Rosemeyer!*, pp. 110–18.

"What are my plans": L'Intransigent, October 15, 1936.

"season of rumors": Motorsport, December 1936.

150   *"most beautiful of the beauties":* Tissot, *Figoni Delahaye*, p. 55.

"I thought, ambitious": Dreyfus and Kimes, *My Two Lives*, p. 69.

As the weeks passed: Ibid., pp. 70–71.

Jean-Pierre Wimille: L'Auto, October 20, 1936.

On December 10, 1936: L'Intransigent, December 11, 1936.

151   *"The story of racing": L'Auto*, December 11, 1936.

152 *"Million Franc Race"*: L'Auto, January 1, 1937.

"I will try my luck": L'Auto, January 6–15, 1937.

"It was to catch the imagination": Blight, *The French Sports Car Revolution*, p. 289.

in mid-January 1937: L'Auto, February 3, 1937.

Initially René was reluctant: Dreyfus and Kimes, *My Two Lives*, p. 70.

153 *In Hamburg, René took the wheel*: La Vie Automobile, February 25, 1937; *Autocar*, February 5, 1937; *Motor*, February 2, 1937; *Motorsport*, February 1937.

155 *"Which way do you want to turn"*: Dreyfus and Kimes, *My Two Lives*, p. 71.

"Bet that today": L'Intransigent, January 24–31, 1937.

156 *René agreed*: L'Auto, January 24–February 3, 1937; *La Vie Automobile*, February 25, 1937.

Lucy was well versed: Lucy Schell, letter to Maurice Phillipe, May 8, 1938, Maurice Phillipe Papers, REVS.

the chaine maudite *(the damned chain)*: Dreyfus and Kimes, *My Two Lives*, p. 71.

In February 1937: Car and Driver, November 1985.

In only six months: Nixon, *Racing the Silver Arrows*, pp. 178–81. In his fine history of the Silver Arrows, Nixon includes a wonderful interview with Uhlenhaut.

157 *To begin with*: Ibid.; Scheller and Pollak, *Rudolf Uhlenhaut*, pp. 46–53; Ludvigsen, *Mercedes-Benz Racing Cars*, p. 112.

For several weeks: "Programm fur die Proben mit den neuen Rennwagenmodel 1937," February 19, 1937, Mercedes-Benz documents, Personal Papers of Karl Ludvigsen, REVS; Jenkinson, *The Grand Prix Mercedes-Benz*, pp. 16–18.

He reached 88 mph: Undated Motortext press release, Personal Papers of Karl Ludvigsen, REVS.

While Rudi was trialing: L'Auto, April 6, 1937.

The team was headed: Motorsport, March 1937.

158 *Next, René competed*: L'Auto, April 7, 1937; *Autocar*, April 9, 1937.

"past great efforts": L'Auto, March 28, 1937.

When reporters: Blight, *The French Sports Car Revolution*, p. 306.

On April 12: L'Auto, August 28, 1937.

159 *"Bravo, Jean-Pierre"*: L'Auto, April 13, 1937.

Throughout April: Blight, *The French Sports Car Revolution*, p. 316.

9. The Winged Beetle

160 *In May, René found himself:* L'Auto, May 10, 1937.

On weekends: L'Auto, March 16, 1937.

It was over 100 degrees: Moretti, *Grand Prix Tripoli,* pp. 129–33, 145; see the 1937 entry for the Tripoli Grand Prix at *The Golden Era of Grand Prix Racing* (website), http://www.kolumbus.fi/leif.snellman/gp371.htm#9.

161 *"champagne all around":* Motorsport, February 2006.

Jean François kept tinkering: Blight, *The French Sports Car Revolution,* pp. 316, 351.

"I'm leaving Louis": Car and Driver, November 1985.

Further, although René did not believe: Motorsport, March 2005. This Nigel Roebuck interview with Dreyfus offers some of the most revealing insights into the driver's views of his fellow competitors, particularly those from Germany. For instance: "Nuvolari was utterly supreme. He could do things with a car no one else could — you were aware of that when you followed him. Often you went into a corner behind him, and just knew that he wasn't going to make it — but he did. [Rudi] Caracciola perhaps *believed* himself the best, and he was indeed a great driver: smooth and, without doubt, the best in the rain. But he was not Nuvolari."

162 *On June 19:* Rao, *Rudolf Caracciola,* p. 237.

At AVUS on May 30: Lang, *Grand Prix Driver,* pp. 49–51.

"Well done, my dear fellow": Neubauer, *Speed Was My Life,* p. 75.

The race, founded by the wealthy railroad family: Dick, *Auto Racing Comes of Age,* pp. 15–25.

Victory there would boost: Cancellieri, *Auto Union,* p. 90.

During their five-day voyage: Rosemeyer and Nixon, *Rosemeyer!,* p. 135; Chakrabongse, *Dick Seaman,* p. 121; Neubauer, *Speed Was My Life,* p. 98.

163 *"In future, Jews must not be allowed":* Nixon, *Racing the Silver Arrows,* p. 166.

"Nazis!" and threw rotten cabbages at them: Neubauer, *Speed Was My Life,* p. 98.

Such were the crowds: Pinkerton Detective Agency, July 15, 1937, 157/1114, DBA.

"This Rosemeyer family": Rosemeyer and Nixon, *Rosemeyer!,* p. 140.

American champion Rex Mays: Autocar, July 9, 1937.

164 *In the end, Bernd easily won the race:* Neubauer, *Speed Was My Life,* p. 99.

"Sir, I think you're just grand": *Autocar*, July 9, 1937.

"most spectacular automobile": *New York Times*, July 6, 1937.

"swastika flag hung": These quotes and the general propaganda impression left by the win is drawn from Bretz, *Bernd Rosemeyer*, p. 102; Day, *Silberpfeil und Hakenkreuz*, pp. 100–104; "Vanderbilt-Rennen in USA," July 8, 1937, Daimler-Benz Aktiengesellschaft, 157/1114, DBA; and undated news clip from *Motor und Sport*, 1937, 157/1114, DBA.

"We hope that you win!": undated news clip from *Motor und Sport*, 1937, 157/1114, DBA.

Mercedes and Auto Union drivers returned to Berlin: Rosemeyer and Nixon, *Rosemeyer!*, pp. 144–45.

165   *"as if this were a Grand Prix"*: Dreyfus and Kimes, *My Two Lives*, p. 76; *Motor*, June 22, 1937; *Motorsport*, July 1991.

"In its place": Blight, *The French Sports Car Revolution*, pp. 351–52.

"pug-nosed dumdum bullet": Ibid., p. 352; *Classic and Sports Car*, undated news clip, René Dreyfus Scrapbooks, MMA.

166   *"How is she going to behave?"*: *L'Intransigent*, June 27, 1937.

There was no deafening yowl: Strother MacMinn, "Delahaye Type 145 Coupe," PPRA; *Classic and Sports Car*, undated news clip, René Dreyfus Scrapbooks, MMA.

167   *"Some countries would like"*: *L'Auto*, March 23, 1937.

"winged beetle": *Autocar*, July 9, 1937.

168   *Since its first test run*: Dreyfus, interview with Caron, 1973.

"Obviously, it would be miraculous": *Le Journal*, July 5, 1937.

By the second lap: Ibid.; *L'Auto*, July 5, 1937; *Autocar*, July 9, 1937; *Autocar*, July 1937.

René wondered again: Blight, *The French Sports Car Revolution*, p. 365.

In a postrace interview: *L'Auto*, June 26, 1937.

on the triangular circuit: *L'Auto*, July 15, 1937; *L'Auto*, July 19, 1937; Blight, *The French Sports Car Revolution*, p. 371.

169   *"Shut up, René"*: Dreyfus and Kimes, *My Two Lives*, p. 77.

The accolades: *Motorsport*, November 2001.

170   *"the whole of sporting Germany"*: *Autocar*, July 30, 1937.

"cream of the European drivers": *Autocar*, July 23, 1937.

"Make a perfect start": *Motor Trend*, April 1975.

On green, he stabbed the pedals: Monkhouse, *Motor Racing with Mercedes Benz*, pp. 57–76; *Motorsport*, August 1937.

171 *"Now, I must ask you"*: Caracciola, *A Racing Car Driver's World*, pp. 163–64.

Production figures at Mercedes: Pohl, Habeth-Allhorn, and Brüninghaus, *Die Daimler-Benz AG in den Jahren*, pp. 134–47.

The national autobahn project: Taylor, *Hitler's Engineers*, pp. 1–40.

172 *"Caracciola, the man without nerves"*: Day, *Silberpfeil und Hakenkreuz*, pp. 157–59.

He published a best-selling memoir: Caracciola, *Rennen*, pp. 1–30.

"The driver fights": Hochstetter, *Motorisierung und "Volksgemeinshaft,"* pp. 298–99.

"racetrack battle": Reuss, *Hitler's Motor Racing Battles*, p. 29; Day, *Silberpfeil und Hakenkreuz*, pp. 148–50.

10. "Le Drame du Million"

173 *In late July:* René Dreyfus, "Ma Course Au Million," reprinted in *Delahaye Club Bulletin*, June 2011.

As August progressed: L'Auto, August 1–23, 1937.

174 *"In bars and cafés"*: Blight, *The French Sports Car Revolution*, p. 377.

The Germans, notably engineers: Undated, unsourced interview with René Dreyfus, *Automobile Quarterly*.

"The engine turned": Dreyfus, interview with Caron, 1973.

175 *"It's very simple"*: Ibid.

With each day of tinkering: Sports Car Guide, September 1959; Dreyfus, "Ma Course Au Million."

The sinuous circuit: Lyndon, *Grand Prix*, pp. 163–64; Boddy, *Montlhéry*, p. 87.

Its 12.5 kilometers of track and road: Motor, December 22, 1954; Blight, *The French Sports Car Revolution*, p. 353.

177 *René brought down:* L'Auto, August 10–18, 1937.

This was a dramatic reduction: L'Intransigent, July 4–6, 1937.

Bugatti announced: L'Auto, August 8–14, 1937.

"boldness and mastery": L'Auto, June 26, 1937.

In the south of France: L'Auto, August 15, 1937; Paris and Mearns, *Jean-Pierre Wimille*, pp. 118–19; Blight, *The French Sports Car Revolution*, pp. 379–80.

179 *"Bugatti will have to fly"*: L'Auto, August 18, 1937.

"Grievous Wounds": undated news article, as quoted in Paris and Mearns, *Jean-Pierre Wimille*, p. 119.

"Ah, if I was ten years younger": *L'Auto,* August 24, 1937.

180 *"It's possible"*: Ibid.

"All other armies": Weber, *The Hollow Years,* p. 251.

"Above all, no war": Ibid., p. 23.

That year alone: "Bericht uber Prufung der im Geshaftsjahr 1937," 12.26, DBA; Jolly, *Delahaye: Sport et Prestige,* pp. 110–12.

181 *"national interest"*: "Protokoll uber die am Donnestag, den 28. Juli 1937," board minutes, 1×01 0021, DBA.

This belied how: Daley, *Cars at Speed,* p. 210.

Tax relief, busted trade unions: Gregor, *Daimler-Benz in the Third Reich,* pp. 36–38.

With board member Jakob Werlin: Ibid., pp. 61–70.

They had begun planning: "Mercedes-Benz 3-Liter Grand Prix Car — Report," 3069/1, DBA; Ludvigsen, *Mercedes-Benz Racing Cars,* pp. 167–72; Ludvigsen handwritten notes on the W154, Personal Papers of Karl Ludvigsen, REVS.

After more development: "Protokoll uber die am Donnestag, den 28. Juli 1937," board minutes, 1×01 0021, DBA.

In August: Alfred Neubauer, "Betrifft: Internationale Rennformel 1938–1940," 170/1136/1, DBA.

182 *Monsieur Charles:* Undated, unsourced interview with René Dreyfus, René Dreyfus Scrapbooks, MMA.

To increase its speed: Report on chassis #48771, PPRA.

Competing for the Million: René Dreyfus, "Dotation du Fonds de Course," undated and unsourced, PPRA.

183 *The offside rear wheel:* Blight, *The French Sports Car Revolution,* p. 381.

The weather looked fine: *L'Auto,* August 26, 1937.

"Dreyfus, I trust you": Dreyfus, "Ma Course Au Million."

"No, I can't do it": Dreyfus, interview with Caron, 1973; Dreyfus, "Dotation du Fonds de Course"; Jolly, *Delahaye: Sport et Prestige,* pp. 144–46; Dreyfus, fiftieth-anniversary speech, September 1993, *Delahaye Club Bulletin.* Dreyfus recounted this ruse in many ways over the years to many interviewers. This conversation is assembled from these four authoritative sources.

The little dog: *L'Auto,* September 1937.

11. The Duel

186 *René drove out of Paris:* L'Auto, August 27–September 15, 1937;
L'Intransigent, August 29, 1937; Dreyfus, "Ma Course Au Million"; Drey-
fus, interview with Caron, 1973; notes on interview with Dreyfus by J. P.
Bernard. PPRA; Dreyfus, "Dotation du Fonds de Course"; Dreyfus and
Kimes, *My Two Lives,* pp. 81–82; René Dreyfus, letter to Martin Dean,
June 14, 1985, René Dreyfus Scrapbooks, MMA; Blight, *The French Sports
Car Revolution,* pp. 381–82; Jolly, *Delahaye: Sport et Prestige,* pp. 145–47;
Dreyfus, fiftieth-anniversary speech, September 1993, *Delahaye Club Bulletin.* The author drew on scores of primary and secondary sources, includ-
ing photographs from the Mullin Museum and elsewhere, to reenact the
August 27, 1937, Million Franc run by René in fine detail. These are the
chief sources the author used throughout the two sections recounting the
events. Quotes and other critical material are noted separately.
"branchless and petrified trees": Autocar, December 21, 1934.

187 *"You have our confidence":* Dreyfus, "Ma Course Au Million."

188 *At the end of the straight:* Labric, *Robert Benoist,* pp. 162–65. In *Grand Prix,*
Lyndon provides an exquisite description of Montlhéry. Any other
description of the road course is augmented by the author's own visit as
well as a detailed schematic, including distances and gradients provided in
Pascal, *Les Grandes Heures de Montlhéry,* pp. 30–31.

190 *Five minutes, 22.9 seconds:* L'Auto, August 28, 1937. All times and speeds are
derived from tables published by *L'Auto.* To be completely accurate, René
needed to run an average lap speed of five minutes, 7.15 seconds, rather
than simply five minutes, seven seconds, to beat one hour, twenty-one
minutes, 54.4 seconds. Over sixteen laps, this small percentage difference
makes over two seconds of difference in favor of the "deficit" René needed
to win back. To avoid a trial in mathematics, the author rounded down the
figures. Any errors are his alone.
"That's not enough": Dreyfus, "Dotation du Fonds de Course."

191 *Any error or misjudgment:* Motorsport, January 1933.
In the seventh lap: Robert Puval, "Un Million pour Quelques Dixièmes,"
undated, unsourced news clip, René Dreyfus Scrapbooks, MMA; "A 210
Kilometeres a l'Heure René Dreyfus Lache le Volant," undated, unsourced
news clip, René Dreyfus Scrapbooks, MMA. In these two articles drawn

from the Mullin Museum Archives on René Dreyfus, the journalists recount rides on the Montlhéry track (one with René himself) that put the reader into the moment and convey how it felt.

195   *"4.9 Seconds Are Worth 1 Million": L'Auto,* August 28, 1937.
"You understand why": L'Intransigent, August 29, 1937.
Jean-Pierre had arrived: Paris and Mearns, *Jean-Pierre Wimille,* p. 121.
Lucy planned: Blight, *The French Sports Car Revolution,* p. 383.

196   *"Tomorrow morning": L'Auto,* August 29, 1937.
"Since everything": L'Intransigent, August 29–30, 1937.
On the third and guaranteed final day: Dreyfus, "Ma Course Au Million."

197   *She had come to expect such slights:* Lucy Schell, letter to Maurice Phillipe, May 5, 1938, Maurice Phillipe Papers, REVS.
"Could they be conjuring up": Blight, *The French Sports Car Revolution,* p. 383.
Considering that Bugatti: L'Auto, August 31, 1937.
"Delahaye: Temporary Keeper": Ibid.
At 4:00 p.m.: L'Auto, September 1, 1937.

198   *With the sun:* Dreyfus, "Ma Course Au Million"; *Le Journal,* September 1, 1937; *L'Intransigent,* September 1, 1937; Dreyfus, interview with Caron, 1973; Paris and Mearns, *Jean-Pierre Wimille,* pp. 121–23.

199   *"It's too late": L'Auto,* September 1, 1937.
At 7:02 p.m., the single-seater sped away: Le Fanatique de L'Automobile, 1978.

200   *"Bravo!":* Paris and Mearns, *Jean-Pierre Wimille,* preface.
"driven like a god": L'Action Automobile, September 1937; Jolly, *Delahaye V12,* pp. 10–11.
"jewel in the dazzling crown": L'Action Automobile, September 1937.
"creator of Écurie Bleue": Dreyfus, "Ma Course Au Million."
"Will we count you": Letter from the Basco-Béarnais Automobile Club Committee, September 1937, René Dreyfus Scrapbooks, MMA.
Some of the letters: Letters to René Dreyfus, September 1937, René Dreyfus Scrapbooks, MMA.
The money was insignificant: Dreyfus and Kimes, *My Two Lives,* p. 84.

12. "One of Us Will Die"

202   *One morning in October 1937: L'Auto,* October 9, 1937; Larsen, *Talbot-Lago Grand Sport,* pp. xvi–xvii; *Autocar,* October 8, 1937.

203 *"Our industry French automotive industry"*: *Le Figaro,* October 13, 1937.

The speakers at the ACF's annual dinner: *L'Auto,* October 16, 1937.

"The Paris Show": "Karosserie-Bericht Uber die Automobil-Ausstellung
—Paris 1937," 12.6, DBA; "Bericht uber den Besuch der Pariser Automo-
bileausstellung 1937," 12.6, DBA.

"Not too impressed": *Motor und Sport,* September 11, 1937; Blight, *The French
Sports Car Revolution,* p. 385.

"imminent danger": *L'Auto,* October 22, 1937.

At the airport outside Frankfurt: *Autocar,* November 5, 1937.

Built to break records: Ludvigsen, *Mercedes-Benz Racing Cars,* pp. 140–43.

204 *Since motor cars were invented*: Howe, *Motor Racing,* p. 226.

There were many records: Ibid., pp. 226–27.

From the moment: Reuss, *Hitler's Motor Racing Battles,* pp. 57–61.

As early as 1934: *L'Auto,* March 7, 1934.

In this way: Dr. Kissel, letter to Gerhart Naumann, May 22, 1937, 11.2,
DBA.

With its long stretches: Nixon, *Racing the Silver Arrows,* p. 157.

The following year: Rosemeyer and Nixon, *Rosemeyer!,* pp. 132–33.

205 *The annual event*: Day, *Silberpfeil und Hakenkreuz,* pp. 126–27.

Newsreel cameramen and photographers: Ludvigsen, *Mercedes-Benz Racing Cars,*
p. 144; *Autocar,* November 5, 1937. In attendance, *Autocar's* John Dugdale
provides fascinating insight into the aerodynamic engineering of Mercedes. See also his book *Great Motor Sport of the Thirties.*

Since first joining Auto Union: Nixon, *Racing the Silver Arrows,* p. 206.

Every seam in the road: Rosemeyer and Nixon, *Rosemeyer!,* p. 168.

206 *"crossing the Niagara Falls on a tightrope"*: Ibid., p. 208.

a son, Bernd Jr.: Ibid., p. 170.

"We proved once again": *Motor und Sport,* October 31, 1937, and November 7,
1937.

"We cannot go on this way": Nolan, *Men of Thunder,* p. 180.

Three days after: *L'Intransigent,* October 29, 1937.

207 *A few weeks later*: *L'Auto,* November 20, 1937; *Autocar,* November 26, 1937.

"Business done": *L'Auto,* November 30, 1937.

"We shall soon": *Motorsport,* December 1937.

On the French side: Ibid.; *Motor und Sport,* December 12, 1937.

208 *"I fancy I can see"*: Court, *A History of Grand Prix Motor Racing,* p. 248.

As for Maserati: Motor, March 8, 1938.

"Going on past form": Autocar, November 19, 1937.

Shortly before Christmas 1937: La Vie Automobile, January 10, 1938; *L'Auto,* December 21, 1937.

209 *In Frankfurt, on January 27, 1938:* Rao, *Rudolf Caracciola,* p. 288.

"I'm convinced": Rosemeyer and Nixon, *Rosemeyer!,* pp. 178–79.

213 *For a few minutes:* Ludvigsen, *Mercedes-Benz Racing Cars,* pp. 151–53; Neubauer, *Speed Was My Life,* pp. 106–9; Caracciola, *A Racing Car Driver's World,* pp. 121–25; Rao, *Rudolf Caracciola,* pp. 289–90; *Motor und Sport,* February 6, 1938. This description of Caracciola's record run is particularly indebted to his memoir and the account by Neubauer. All quotes and details come from the sources noted.

By nine o'clock: "Bericht Rekorde Frankfurt/Main," February 2, 1938, 428/3020, DBA; Aldo Zana, "A Roadmap for a Tentative Explanation of Bernd Rosemeyer's January 28, 1938 Accident," *The Golden Era of Grand Prix Racing* (website), http://www.kolumbus.fi/leif.snellman/zana.htm; Caracciola, *A Racing Car Driver's World,* pp. 125–28; Neubauer, *Speed Was My Life,* pp. 110–14; Rosemeyer and Nixon, *Rosemeyer!,* pp. 179–82. Feuereissen wrote an evocative account of Rosemeyer's crash in Nixon's book. The Zana study is the most authoritative on what exactly happened that fateful January day. All quotes and details come from the sources noted here.

216 *"heavy day":* Goebbels, *Die Tagebücher,* pp. 121–22.

Goebbels perceived an opportunity: Zana, "A Roadmap for a Tentative Explanation of Bernd Rosemeyer's January 28, 1938 Accident."

"the news of your husband's tragic fate": Frilling, *Elly Beinhorn und Bernd Rosemeyer,* pp. 321–22.

"He died as a soldier": Ibid.

"Bernd Rosemeyer—friend": Day, *Silberpfeil und Hakenkreuz,* pp. 185–86.

The tragedy did not stop Rudi: "Niederschrift betreffend Vorstandsitzung vom 8. Marz 1938 in Untertürkheim, Kissel Protokolle 1938," 0026, I/11, DBA.

217 *"Fate took him from us": Motor* (Deutschland), March 8, 1938.

"chivalrous match": "Betr.: Rekordversuch," note between Hühnlein and Kissel, December 1, 1937, 12/25, DBA.

13. "Find Something"

218 *On January 19, 1938: Delahaye Club Bulletin,* March 2002.

After an extended: Dreyfus and Kimes, *My Two Lives,* p. 84.

Crossing the Mediterranean: Ibid.; *Delahaye Club Bulletin,* March 2002; *L'Auto,* January 25–30, 1938.

219 *"It's a tragic accident": L'Auto,* January 29, 1938.

René did not care to dwell on it: Ribet, interview with the author.

"Once the horror": Daley, *Cars at Speed,* p. 187.

220 *He and Laury:* Blight, *The French Sports Car Revolution,* p. 421; *Le Fanatique de L'Automobile,* April 1978.

Jean François had managed: Jolly, *Delahaye V12,* p. 12.

Instead, he believed: "A Celebration of the Life of René Dreyfus," PPDF.

On February 18: Autocar, February 18–25; *La Vie Automobile,* March 10, 1938; *Motor und Sport,* February 27, 1938; Chakrabongse, *Dick Seaman,* p. 136.

Since early February: Kershaw, *Hitler,* pp. 56–76.

221 *"You don't believe":* Ibid., p. 70.

"the end of Austria": Shirer, *Berlin Diary,* p. 92.

"beloved child": L'Auto, February 19–22, 1938.

"When I had the privilege": Motor und Sport, February 27, 1938; Domarus, *Hitler,* pp. 1018–20.

"fellowship between the Führer and his listeners": L'Auto, February 19–22, 1938.

"the best and most courageous": Motor, March 1, 1938; Domarus, *Hitler,* pp. 1018–20.

Behind him, curtains were swept aside: Chakrabongse, *Dick Seaman,* p. 137; *Motorsport,* March 1938; *Motor Italia,* March 1938.

222 *"over ten million Germans":* Kershaw, *Hitler,* p. 73.

"a free, independent": Shirer, *The Rise and Fall of the Third Reich,* p. 334.

"a state of ecstasy": Ibid., p. 334.

"Individual Jews were robbed": Kershaw, *Hitler,* p. 84.

223 *"The hard fact":* Shirer, *Berlin Diary,* p. 107.

"Britain and France": Ibid.

In a park somewhere outside Milan: Motor und Sport, March 27, 1938.

"ten thousand scalded cats": Motorsport, April 1938.

Since early March: Autocar, January 29, 1937.

224 *The V12 engine:* "Development of the W154," Ludvigsen notes, Personal Papers of Karl Ludvigsen, REVS; Monkhouse, *Motor Racing with Mercedes Benz,* pp. 31–32.
Every day they drove the car: Chakrabongse, *Dick Seaman,* p. 107.
Everything was noted: "Abschrift: Mercedes Probefahrten in Monza," 183/1162, DBA.
Neubauer and Uhlenhaut: Ludvigsen, *Mercedes-Benz Racing Cars,* p. 130.
"Espionage Department": "Abschrift: Mercedes Probefahrten in Monza," 183/1162, DBA.
225 *Even so, the mood:* Lang, *Grand Prix Driver,* p. 19.
"clever move": Brauchitsch, *Ohne Kampf Kein Siege,* p. 123.
226 *"We racing drivers":* Völkischer Beobachter, March 26, 1938.
"Outstanding": "Abschrift: Letzter Fahrbericht uber die Proben in Monza —Typ 154," Personal Papers of Karl Ludvigsen, REVS.
"The goods," said Seaman: Chakrabongse, *Dick Seaman,* p. 138.
"goes like the bomb": Motor, April 5, 1938.
"big impression": Motor, March 29, 1938.
It was one week: Dreyfus and Kimes, *My Two Lives,* p. 84; L'Intransigent, April 5, 1938; L'Auto, April 3, 1938; Auto Retro, September 13, 1981.
227 *would not be joining the team:* L'Automobile sur la Côte d'Azur, April 1938.
In money alone: Paris Soir, May 1, 1938.
228 *"Every blow to Schmeling's eyes":* Frilling, *Elly Beinhorn und Bernd Rosemeyer,* p. 101.
Whatever Lucy said to him: Evelyne Dreyfus, interview with the author; Dreyfus and Kimes, *My Two Lives,* p. 87; Ribet, interview with the author.
"find something to struggle and fight for": Sports Car Guide, September 1959; Dreyfus and Kimes, *My Two Lives,* p. 51; Automobile Quarterly, Summer 1967.
As he departed: Auto Moto Retro, October 1987.

14. The Dress Rehearsal

229 *In the nineteenth century, wealthy visitors:* Tucoo-Chala, *Histoire de Pau,* pp. 1–4.
230 *Pau also became known:* Darmendrail, *Le Grand Prix de Pau,* pp. 6–8, 25–27.
Over its 1.7 miles: Helck, *Great Auto Races,* p. 218.

When René arrived: Le Patriote, April 6–8, 1938, ACBB Scrapbook.

In a convoy: L'Auto, April 6, 1938; Earl, *Quicksilver,* pp. 64–65.

231 *"full dress rehearsal":* undated article, *Hamburger Anzeiger,* DBA.

Since the Million run: Blight, *The French Sports Car Revolution,* p. 385.

The Delahayes drew only a fraction: "Betreff: Grosser Preis von Pau, April 8, 1938," 1137/1, DBA.

"All the colors": Le Patriote, March 29, 1938.

Wimille was on the roster: La Dépêche, April 1–8, 1938; *Independent,* April 1–8, 1938.

232 *"surpass in success": L'Auto,* April 8, 1938.

"The World's Greatest Driver": Le Patriote, April 4, 1938.

After a few spins around the course: "Betreff: Grosser Preis von Pau, April 8, 1938," 1137/1, DBA.

Wanting to test: L'Auto Italiana, April 20, 1938; *Il Littoriale,* April 9, 1938; Moretti, *When Nuvolari Raced . . . ,* p. 56.

234 *"the car into the lake": Automobile Quarterly,* 1st quarter 1937.

Such was the confidence: Lang, *Grand Prix Driver,* p. 67.

"high-legged" Delahaye: "Betreff: Grosser Preis von Pau, April 8, 1938," 1137/1, DBA.

Coming out of some corners: Automobile Quarterly, Summer 1964.

Rudi finished the second session: "Betreff: Grosser Preis von Pau, April 9, 1938," 1137/1, DBA; *L'Auto,* April 10, 1938.

235 *Nothing could be left to chance:* undated article in *Speed,* Papers of Karl Ludvigsen, REVS; Kimes, *The Star and the Laurel,* p. 251.

At the end of the day: Walkerly, *Grand Prix,* pp. 40–41.

Lucy was certain that Delahaye: L'Auto, April 30–May 5, 1938; Blight, *The French Sports Car Revolution,* pp. 454–55.

236 *"racing things": Automobile Quarterly,* Summer 1964.

"This takes away": "Betreff: Grosser Preis von Pau, April 9, 1938," 1137/1, DBA; "Telefongesprach Director Sailor aus Pau," 18.40 Uhr. 9.4.38, 1137/1, DBA.

237 *"mastery of German engineers":* "Mercedes-Benz Pressedienst, April 5, 1938," 1137/1, DBA.

"defeated the very latest thing": Court, *A History of Grand Prix Motor Racing,* p. 250.

"Surely, Mercedes should win": L'Auto, April 10, 1938.

"The outcome of this race": Automobile Quarterly, Summer 1964.

"beloved leader": Wilhelm Kissel, "Geleitwort zur Wahlhandlung am 10. April 1938," March 31, 1938, 12.13, DBA.

"Hitler led the German automobile industry": Wilhelm Kissel, "Deutschlands Automobile-Industrie dankt dem Führer," March 30, 1938, 12.13, DBA.

238   *"How different has our racing success"*: Robert Hammelehle, "Was denken wir vom Führer und seinem Werk, und warum sagen wir bei der bevorstehenden Wahl: Ja?," March 31, 1938, 12.13, DBA.

"victory in the Grand Prix": Wilhelm Kissel, "Deutschlands Automobile-Industrie dankt dem Führer," March 30, 1938, 12.13, DBA.

"So you think": Dreyfus and Kimes, My Two Lives, p. 85.

Later that night: L'Action Automobile, May 1938; Dreyfus, interview with Caron, 1973; Automobile Quarterly, Summer 1964; Road and Track, December 1988; Darmendrail, Le Grand Prix de Pau, preface by René Dreyfus.

15. Victory at Pau

241   *The dawn sun*: L'Actualité Automobile, May 1938.

Slowly, but with a gathering momentum: Collection of news articles, ACBB Scrapbook; Walkerly, Grand Prix, pp. 36–38. Walkerly provides a brilliant description of the start of a Grand Prix race in the 1930s.

"It was as if the multitude": Ibid., p. 37.

242   *That morning*: "Telefonbericht am 10.4.38," 17.05 Uhr. 1137/1, DBA.

"Are you in agreement": Domarus, Hitler, p. 1089.

Spectators who wanted: Collection of news articles, ACBB Scrapbook. The hundreds of articles from French, German, Italian, and British newspapers that the ACBB collected in an oversized scrapbook for the Pau 1938 race provide much of the color in this chapter.

Rudi ate an underdone steak: Automobile Quarterly, Summer 1968; Neubauer, Speed Was My Life, p. 120.

His teammate Lang: Brauchitsch, Ohne Kampf Kein Siege, p. 120.

In his room at the Hôtel de France: Ribet, interview with the author.

243   *By the grandstands*: Walkerly, Grand Prix, pp. 37–38; Neubauer, Speed Was My Life, pp. 120–21.

One of the nine cars: "Telefongesprach H. Sailer aus Pau 10.4.38," 18.40 Uhr. 1137/1, DBA.

From the original sixteen: L'Action Automobile, May 1938.

In the Delahaye pit: L'Action Automobile, May 1938.

244   *Laury then drew René aside:* Road and Track, December 1988.

The Mercedes driver: Automobile Quarterly, Summer 1968.

As 2:00 p.m. fast approached: See the 1938 entry for the Pau Grand Prix at *The Golden Era of Grand Prix Racing* (website), http://www.kolumbus.fi/leif. snellman/gp381.htm#2.

After the briefing: Speed, January 1938.

"To your cars": Motorsport, June 1938.

Rudi limped: Motor Trend, April 1975.

245   *"My eyes fill with tears":* Automobile Quarterly, Summer 1968.

In the grandstands: Walkerly, *Grand Prix,* p. 38.

"various dopes": Motorsport, June 1938.

Rudi rechecked his gear level: Motor Trend, April 1975.

Five . . . four . . . three . . . two . . . one!: L'Auto, April 11, 1938; film of the reportings of *Éclair Journal,* Pau 1938, Gaumont-Pathe Archives; collection of news articles, ACBB Scrapbook; Dreyfus and Kimes, *My Two Lives,* pp. 85–87; *Motorsport,* May 1938; *Road and Track,* December 1988; *Paris Soir,* April 11, 1938; Darmendrail, *Le Grand Prix de Pau,* pp. 57–64; "Telefonanruf H. Kudorfer 10.4.38," 15–18 Uhr. 1137/1, DBA; Lang, *Grand Prix Driver,* pp. 67–68; *Automobile Quarterly,* Summer 1964. Details and color on the Pau Grand Prix 1938 were largely drawn from these sources. Any quotes or important details unique to a particular source are cited separately. Of especially good use were the internal Mercedes-Benz reports and the ACBB Scrapbook.

247   *"paralyzing wheelspin": Road and Track,* September 1988.

In that, a racing driver: Hilton, *Inside the Mind of the Grand Prix Driver,* p. 124.

René followed the Avenue du Général Poeymirau: "Programme, Grand Prix Automobile, Pau 1938," 170/1137/1, DBA.

248   *By the time he hit 100 mph:* ACBB, "Pau: Grand Prix de Vitesse: Report," 170/1137/1 1938, DBA. All lap times in the race are drawn from this ACBB report distributed to the press.

As René had experienced in the past: "Haus-Pressedienst der Daimler-Benz," April 16, 1938, 1137/1, DBA.

252   *"Caracciola and Dreyfus":* "Telefonanruf H. Kudorfer 10.4.38," 15 Uhr. 1137/1, DBA.

Among them was a car-crazy fourteen-year-old Jewish kid: Automobile Quarterly, 2nd quarter 1980.

"scientific driver": Moteurs Course, 3rd trimester 1956.

254 *He was expected:* Hochstetter, *Motorisierung und "Volksgemeinshaft,"* p. 293. This is almost a direct quote from a Caracciola memoir.

255 *"There is little chance":* L'Auto, April 11, 1938.

"It seems very doubtful": "Telefonanruf H. Kudorfer 10.4.38," 16:05 Uhr. 1137/1, DBA.

Epilogue

259 *"surpassed all my expectations":* Paris Soir, April 11, 1938; *New York Times,* April 11, 1938.

That same night in Pau: Le Patriote, April 13, 1938.

"right day, the right driver": Jolly, *Delahaye V12,* pp. 12–13.

"One cannot congratulate": La Vie Automobile, April 25, 1938.

"The success achieved": Paris Soir, April 11, 1938.

"A Beautiful French Victory!": Le Figaro, April 11, 1938; L'Auto, April 11, 1938.

260 *"There was something":* Automobile Quarterly, Summer 1964; *Motor,* April 19, 1938.

"The first race": Motorsport, May 1938.

"dress rehearsal": "Telefonanruf H. Kudorfer 10.4.38," 18:30 Uhr. 1137/1, DBA; "Telefonanruf H. Kudorfer 10.4.38," 20:15 Uhr. 1137/1, DBA; German newspapers, as quoted in Herzog, *Unter dem Mercedes-Stern,* pp. 121–22.

"an embarrassing defeat": Nye and Goddard, *Dick and George: The Seaman-Monkhouse Letters,* p. 194.

"Phantastique": Dreyfus and Kimes, *My Two Lives,* p. 87.

Two weeks later: Chakrabongse, *Road Start Hat Trick,* pp. 162–65; *Motor,* April 26, 1938.

"The first year": Autocar, April 29, 1938.

Jean François promised: L'Intransigent, April 25, 1938; Dreyfus, interview with Caron, 1973.

261 *"We always tend to imagine":* Lucy Schell, letter to Maurice Phillipe, May 5, 1938, Maurice Phillipe Papers, REVS.

"Please take note": L'Auto, April 28–May 5, 1938; Blight, *The French Sports Car Revolution,* 456.

After Cork: Moretti, *Grand Prix Tripoli,* pp. 143–45.

262 *"divine avenger":* Automobile Quarterly, 2nd quarter 1980; *New York Times,* January 30, 1979.

In a systematic orgy of terror: Wolff, *The Shrinking Circle,* pp. 62–65.

Lucy was distracted: Dreyfus and Kimes, *My Two Lives,* p. 91; *Le Journal,* September 20, 1938.

"Vive la France!": Dreyfus and Kimes, *My Two Lives,* p. 93.

263 *Less than a month later:* Le Journal, September 20, 1938.

On leave from the French transport corps: Dreyfus and Kimes, *My Two Lives,* pp. 101–20.

264 *"Dreyfus was to the French people":* Ibid., p. 113; *Auto Age,* August 1956; *Autosport,* March 1, 1955.

265 *When war broke out:* Caracciola, *A Racing Car Driver's World,* pp. 166–68; Neubauer, *Speed Was My Life,* pp. 166–69; Molter, *German Racing Cars and Drivers,* p. 7.

During the war: Gregor, *Daimler-Benz in the Third Reich,* pp. 50–96; Bellon, *Mercedes in Peace and War,* pp. 233–54.

266 *Nur Säufer, keine Kämpfer:* Hilton, *Grand Prix Century,* p. 111.

NSKK troops proved: Hochstetter, *Motorisierung und "Volksgemeinshaft,"* pp. 421–54.

"spearhead of the modern war": Ibid., p. 426.

At the end of the war: Molter, *German Racing Cars and Drivers,* p. 7.

After Germany's surrender: Caracciola, *A Racing Car Driver's World,* pp. 180–96.

267 *During that time:* Motorsport, January 1999.

In 1952, Neubauer enlisted Rudi: Caracciola, *A Racing Car Driver's World,* pp. 202–14.

"There is no longer any object": Cernuschi, *Nuvolari,* p. 178.

268 CORRERAI PIU FORTE PER LE VIE DEL CIELO: Moretti, *When Nuvolari Raced . . . ,* p. 54.

"constructive non-cooperation": Automobile Quarterly, July 1999.

not alone within the French racing community: See Saward, *The Grand Prix Saboteurs,* for a fascinating read on the French Resistance efforts of Benoist and his compatriots.

In September 1945: Hilton, *Grand Prix Century,* pp. 194–95.

269 *"the undisputed star of the salons":* J. P. Bernard, "The History of Delahaye," essay in PPRA.

the company struggled to survive: Automobile Quarterly, July 1999.

Twice a week: Original interviews between Dreyfus and Kimes, PPBK; Dreyfus and Kimes, *My Two Lives,* pp. 139–40.

By early evening: Autosport, March 11, 1955.

270 *"I loved racing": Sports Car Graphic,* September 1984.

"I am listed": Dreyfus and Kimes, *My Two Lives,* p. 54.

In 1993: Evelyne Dreyfus, interview with the author; Philip Dreyfus, speech at the funeral of René Dreyfus, PPBK.

"I've lived so many things": Ribet, interview with the author.

271 *There were rumors that Hitler:* Peter Mullin, interview with the author, Los Angeles, 2017.

The initial fates of the four Écurie Bleue 145s: Adatto and Meredith, *Delahaye Styling and Design,* pp. 257–307; André Vaucourt, reports on the provenance of the Delahaye 145s.

"apex of the automobile world": Peter Mullin, interview with the author.

"most beautiful cars in the world": André Vaucourt, article on the Delahaye V12, chassis # 48771, sent to the author.

272 *"rolling sculpture":* Sam Mann, interview with the author, New Jersey, 2019.

The last of the four Delahaye 145s: Ibid.; Delahaye 145 in PPRA; correspondence between André Vaucourt and the author, 2018. It should be noted that homologation experts believe that the original chassis number assigned by the factory is the key to the puzzle, since neither of the chassis have suffered much alteration over the years. But if Delahaye or Lucy Schell maintained records about which chassis raced with which engine and body, at which race, they have been lost. Further, parts were often swapped between cars, and teams frequently doctored numbers and plates to avoid custom duties when traveling to races outside France.

Sources and Bibliography

Archival and Personal Papers

Automobile Club Basco-Béarnais, France
Automobile Club de France, France
Automobile Club de Nice, France
Mullin Museum Archives, United States
Simeone Foundation Museum Archive
Daimler-Benz Archive, Germany
Revs Institute, United States
Bundesarchiv, Germany
Cord Duesenberg Library, United States
Archives de Châtel-Guyon, France
Archives de Brunoy, France
Personal Papers of Richard Adatto
Personal Papers of the Dreyfus Family
Personal Papers of Beverly Kimes, Cord Duesenberg Library
Personal Papers of Maurice Louche
Personal Papers of Karl Ludvigsen
Personal Papers of Maurice Phillipe
Personal Papers of Anthony Blight

Books

Abeillon, Pierre, *Talbot-Lago de Course* (Paris: Vandoeuvres, 1992).

Adatto, Richard, *From Passion to Perfection: The Story of French Streamlined Styling, 1930–39* (Paris: Éditions SPE Barthelemey, 2003).

———. *French Curves* (Oxnard, CA: Mullin Automotive Museum, 2011).

Adatto, Richard, and Diana Meredith, *Delahaye Styling and Design* (Philadelphia: Coachbuilt Press, 2006).

Beadle, Tony, *Delahaye: Road Test Portfolio* (Surrey, UK: Brooklands Books, 2010).

Belle, Serge, *Blue Blood: A History of Grand Prix Racing Cars in France* (London: Frederick Warne, 1979).

Bellon, Bernard, *Mercedes in Peace and War: German Automobile Workers, 1903–45* (New York: Columbia University Press, 1990).

Birabongse, Prince of Thailand, *Bits and Pieces* (London: Furnell and Sons, 1942).

Birkin, Henry, *Full Throttle* (London: G. T. Foulis & Co., 1945).

Blight, Anthony, *The French Sports Car Revolution: Bugatti, Delage, Delahaye and Talbot in Competition 1934–1939* (Somerset, UK: G. T. Foulis & Co., 1990).

Blumenson, Martin, *The Vilde Affair: Beginnings of the French Resistance* (Boston: Houghton Mifflin, 1977).

Boddy, William, *Montlhéry: The Story of the Paris Autodrome* (Dorchester, UK: Veloce Publishing, 2006).

Bouzanquet, Jean François, *Fast Ladies: Female Racing Drivers from 1888 to 1970* (London: Veloce, 2009).

Bradley, W. F., *Ettore Bugatti* (Abingdon, UK: Motor Racing Publications, 1948).

Brauchitsch, Manfred von, *Ohne Kampf Kein Siege* (Berlin: Verlag der Nation, 1966).

Bretz, Hans, *Bernd Rosemeyer: Ein Leben Für den Deutschen Sport* (Berlin: Wilhelm Limpert-Verlag, 1938).

———. *Mannschaft und Meisterschaft: Eine Bilanz der Grand Prix Formel 1934–37* (Stuttgart: Daimler Benz AG, 1938).

Bugatti, L'Ebe, *The Bugatti Story* (Philadelphia: Chilton Book Co., 1967).

Bullock, Alan, *Hitler: A Study in Tyranny* (New York: Bantam Books, 1958).

Bullock, John, *Fast Women* (London: Robson Books, 2002).

Burden, Hamilton, *The Nuremberg Rallies: 1923–39* (London: Pall Mall Press, 1967).

Cancellieri, Gianna, *Auto Union — Die Grossen Rennen 1934–39* (Hanover: Schroeder and Weise, 1939).

Canestrini, Giovanni, *Uomini E Motori* (Monza: Nuova Massimo, 1957).

Caracciola, Rudolf, *Rennen – Sieg – Rekorde!* (Stuttgart: Union Deutsche Verlagsgesellschaft, 1943).

———. *A Racing Car Driver's World* (New York: Farrar, Straus and Giroux, 1961).

Carter, Bruce, *Nuvolari and Alfa Romeo* (New York: Coward McCann, 1968).

Cernuschi, Giovanni, *Corse per Il Mondo* (Milan: Editoriale Sportiva, 1947).

Cernuschi, Count Giovanni, *Nuvolari* (New York: William Morrow, 1960).

Chakrabongse, Prince Chula, *Road Start Hat Trick: Being an Account of Two Season of "B. Bira" the Racing Motorist in 1937 and 1938* (London: G. T. Foulis & Co., 1944).

———. *Dick Seaman: A Racing Champion* (Los Angeles: Floyd Clymer, 1948).

Cholmondeley-Tapper, Thomas Pitt, *Amateur Racing Driver* (London: G. T. Foulis & Co., 1966).

Cohin, Edmond, *Historique de la Course Automobile* (Paris: Éditions Lariviere, 1966).

Court, William, *A History of Grand Prix Motor Racing, 1906–1951* (London: Macdonald, 1966).

Daley, Robert, *Cars at Speed* (New York: Collier Books, 1961).

Darmendrail, Pierre, *Le Grand Prix de Pau: 1899–1960* (Paris: La Librairie du Collectionneur, 1992).

Day, Uwe, *Silberpfeil und Hakenkreuz* (Berlin: Bebra Verlag, 2005).

Dick, Robert, *Auto Racing Comes of Age: A Transatlantic View of Cars, Drivers and Speedways, 1900–1925* (London: McFarland & Co., 2013).

Domarus, Max, *Hitler: Speeches and Proclamations* (Mundelein, IL: Bolchazy-Carducci Publishers, 1990).

Dorizon, Jacques, François Peigney, and Jean-Pierre Dauliac, *Delahaye: Le Grand Livre* (Paris: EPA Éditions, 1995).

Doyle, Gary, *Carlo Demand in Motion and Color* (Boston: Racemaker Press, 2007).

Dreyfus, René, and Beverly Rae Kimes, *My Two Lives: Race Driver to Restaurateur* (Tucson: Aztec Corp., 1983).

Dugdale, John, *Great Motor Sport of the Thirties: A Personal Account by John Dugdale* (London: Wilton House Gentry, 1977).

Earl, Cameron, *Quicksilver* (London: HMSO, 1996).

Ferrari, Enzo, *My Terrible Joys* (London: Hamish Hamilton, 1963).

François-Poncet, André, *The Fateful Years: Memoirs of a French Ambassador in Berlin, 1931–38* (New York: Howard Fertig, 1972).

Frilling, Christoph, *Elly Beinhorn und Bernd Rosemeyer: Kleiner Grenzverkehr zwischen Resistenz und Kumpanei im Nationalsozialismus* (Frankfurt: Peter Lang, 2009).

Gautier, Jean, and Jean-Pierre Altounian, *Memoire en Images Brunoy* (France: Éditions Alan Sutton, 1996).

Goebbels, Joseph, *Die Tagebücher* (Munich: K. G. Saur Verlag, 2000).

Gregor, Neil, *Daimler-Benz in the Third Reich* (New Haven, CT: Yale University Press, 1998).

Hamilton, Charles, *Leaders and Personalities of the Third Reich* (San Jose, CA: R. James Bender Publishing, 1998).

Helck, Peter, *Great Auto Races* (New York: Harry Abrams, 1975).

Herzog, Bodo, *Unter dem Mercedes-Stern: Die Grosse Zeit der Silberpfeile* (Ernst Gerdes Verlag, 1996).

Hildenbrand, Laura, *Seabiscuit* (New York: Random House, 2001).

Hilton, Christopher, *Hitler's Grands Prix in England: Donington 1937 and 1938* (Somerset, UK: Haynes Publishing, 1999).

———. *Inside the Mind of the Grand Prix Driver* (Somerset, UK: Haynes Publishing, 2001).

———. *Nuvolari* (Derby, UK: Breedon Books Publishing, 2003).

———. *Grand Prix Century: The First 100 Years of the World's Most Glamorous and Dangerous Sport* (Somerset, UK: Haynes Publishing, 2005).

———. *How Hitler Hijacked World Sport: The World Cup, the Olympics, the Heavyweight Championship, and the Grand Prix* (Stroud, UK: History Press, 2012).

Hochstetter, Dorothee, *Motorisierung und "Volksgemeinshaft": Das Nationalsozialistische Kraftfahrkorps (NSKK) 1931–1945* (Munich: R. Oldenbourg Verlag, 2005).

Hodges, David, *The Monaco Grand Prix* (London: Temple Press Books, 1964).

Howe, Lord, *Motor Racing* (London: Seeley Service & Co., 1939).

Jellinek-Mercedes, Guy, *My Father, Mr. Mercedes* (London: Chilton Book Co., 1961).

Jenkinson, Denis, *The Racing Driver* (London: BT Batsford, 1958).

———. *The Grand Prix Mercedes-Benz, Type W125, 1937* (New York: Arco Publishing, 1970).

Jolly, François, *Delahaye V12* (Nimes, France: Éditions du Palmier, 1980).

———. *Delahaye: Sport et Prestige* (Paris: Jacques Grancher, 1981).

Kershaw, Ian, *Hitler: 1936–45* (New York: W. W. Norton, 2000).

Keys, Barbara, *Globalizing Sport: National Rivalry and International Community in the 1930s* (Cambridge, MA: Harvard University Press, 2006).

Kimes, Beverly, *The Star and the Laurel: The Centennial History of Daimler, Mercedes, and Benz* (Montvale, NJ: Mercedes-Benz of North America, 1986).

King, Bob, *The Brescia Bugatti* (Mulgrave, Australia: Images Publishing Group, 2006).

Klemperer, Victor, *I Will Bear Witness: A Diary of the Nazi Years* (New York: Modern Library, 1999).

———. *The Language of the Third Reich: A Philologist's Notebook* (London: Athlone Press, 2000).

Labric, Roger, *Robert Benoist: Champion du Mondo* (Paris: Edicta Paris, 2008).

Lang, Hermann, *Grand Prix Driver* (London: G. T. Foulis & Co., 1953).

Larsen, Peter, with Ben Erickson, *Talbot-Lago Grand Sport: The Car from Paris* (Copenhagen: Dalton, Watson Fine Books, 2012).

Laux, James, *In First Gear: The French Automobile Industry to 1914* (Montreal: McGill-Queen's University Press, 1976).

Lemerie, Jean-Louis, and Emmanuel Piat, *Histoire de l'Automobile Club de France* (Paris: Alcyon Media Groupe, 2012).

Liddell Hart, B. H., *History of the Second World War* (Old Saybrook, CT: Konecky & Konecky, 1970).

Lottman, Herbert, *The Fall of Paris: June 1940* (London: Sinclair-Stevenson, 1992).

Louche, Maurice, *1895–1995: Un Siècle de Grands Pilotes Français* (Nimes, France: Éditions du Palmier, 1995).

———. *Le Rallye Monte Carlo au Xxe Siècle* (Monaco: L'Automobile-Club de Monaco, 2001).

Ludvigsen, Karl, *Mercedes-Benz Racing Cars* (Newport Beach, CA: Bond/Parkhurst Books, 1974).

———. *Classic Grand Prix Cars: The Front-Engined Formula 1 Era 1906–1960* (Gloucestershire: Sutton Publishing, 2000).

———. *The V12 Engine: The Untold Story of the Technology, Evolution, Performance, and Impact of All V12-Engined Cars* (Somerset, UK: Haynes Publishing, 2005).

Lyndon, Barre, *Grand Prix* (London: John Miles, 1935).

Malino, Frances, and Bernard Wasserstein, *The Jews in Modern France* (Lebanon, NH: University Press of New England, 1985).

Marc-Antoine, Colin, *Delahaye 135* (Paris: ETAI, 2003).

Mays, Raymond, *Split Seconds* (London: G. T. Foulis & Co., 1951).

Mitchell, Allan, *Nazi Paris: History of an Occupation, 1940–44* (New York: Berghahn Books, 2010).

Moity, Christian, *Grand Prix de Monaco*, vol. 1 (Besançon: Éditions d'Art/J. P. Barthelemy, 1996).

Molter, Günther, *German Racing Cars and Drivers* (Los Angeles: Floyd Clymer, 1950).

———. *Rudolf "Caratsch" Caracciola: Assergewöhnlicher Rennfahrer und Eiskalter Taktiker* (Stuttgart, Germany: Motor Buch Verlag, 1997).

Mommsen, Hanx, *The Rise and Fall of Weimar Democracy* (Chapel Hill: University of North Carolina Press, 1996).

Monkhouse, George, *Motor Racing with Mercedes Benz* (Los Angeles: Floyd Clymer, 1945).

———. *Grand Prix Racing: Facts and Figures* (London: G. T. Foulis & Co., 1950).

Moretti, Valerio, *Grand Prix Tripoli* (Milan: Automobilia, 1994).

———. *When Nuvolari Raced . . .* (Dorset: Veloce Publishing, 1994).

Neubauer, Alfred, *Speed Was My Life* (New York: Clarkson Potter, 1958).

Neue Deutsche Biographie, vol. 11 (Berlin: Duncker & Humblot, 1977).

Nicholas, Lynn, *The Rape of Europa: The Fate of Europe's Treasures in the Third Reich and the Second World War* (New York: Alfred A. Knopf, 1995).

Nixon, Chris, *Racing the Silver Arrows* (Oxford: Osprey, 1986).

———. *Kings of Nürburgring* (Middlesex, UK: Transport Bookman Publications, 2005).

Nolan, William, *Men of Thunder: Fabled Daredevils of Motor Sport* (New York: G. P. Putnam's Sons, 1964).

Nye, Doug, and Geoffrey Goddard, *Dick and George: The Seaman-Monkhouse Letters, 1936–1939* (London: Palawan Press, 2002).

Orsini, Luigi, and Franco Zagari, *Maserati: Una Storia nell Storia* (Milan: Emmeti Grafica, 1980).

Paris, Jean-Michel, and William Mearns, *Jean-Pierre Wimille: A Bientôt la Revanche* (Paris: Drivers, 2002).

Pascal, Dominique, *Les Grandes Heures de Montlhéry* (Boulogne-Billancourt: ETAI, 2004).

Pohl, Hans, Stephanie Habeth-Allhorn, and Beate Brüninghaus, *Die Daimler–Benz AG in den Jahren 1933 bis 1945* (Stuttgart: Franz Steiner Verlag, 1986).

Pomeroy, Laurence, *The Grand Prix Car: 1906–1939* (Abingdon, UK: Motor Racing Publications, 1949).

Pritchard, Anthony, *Silver Arrows in Camera* (Somerset, UK: Haynes Publishing, 2008).

Purdy, Ken, *The Kings of the Road* (London: Anchor Press, 1957).

Rao, Rino, *Rudolf Caracciola: Una Vita per le Corse* (Bologna, Italy: Edizioni ASI Service, 2015).

Reuss, Eberhard, *Hitler's Motor Racing Battles* (Somerset, UK: Haynes Publishing, 2006).

Rolt, L.T.C., *Horseless Carriage: The Motor-Car in England* (London: Constable Publishers, 1950).

Rosbottom, Ronald, *When Paris Went Dark: The City of Light Under German Occupation, 1940–44* (New York: Little, Brown and Co., 2014).

Rosemann, Ernst, *Um Kilometer und SeKunden* (Stuttgart: Union Deutsche Verlagsgesellschaft, 1938).

Rosemeyer, Elly, and Chris Nixon, *Rosemeyer! A New Biography* (Middlesex, UK: Transport Bookman Publishers, 1986).

Ruesch, Hans, *The Racer* (New York: Ballantine Books, 1953).

Sachs, Harvey, ed., *The Letters of Arturo Toscanini* (New York: Alfred A. Knopf, 2002).

Saward, Joe, *The Grand Prix Saboteurs* (London: Morienval Press, 2006).

Scheller, Wolfgang, and Thomas Pollak, *Rudolf Uhlenhaut: Ingenieur und Gentleman* (Königswinter, Germany: HEEL Verlag, 2015).

Setright, L.J.K., *The Designers: Great Automobiles and the Men Who Made Them* (Chicago: Follett Publishing Co., 1976).

Seymour, Miranda, *Bugatti Queen: In Search of a French Racing Legend* (New York: Random House, 2004).

Shirer, William, *Berlin Diary: The Journal of a Foreign Correspondent 1934–41* (New York: Alfred A. Knopf, 1941).

———. *The Rise and Fall of the Third Reich* (New York: Simon & Schuster, 1990).

Stevenson, Peter, *Driving Forces: Grand Prix Racing Season Caught in the Maelstrom of the Third Reich* (Cambridge: Bentley Publishers, 2000).

Stobbs, William, *Les Grandes Routieres: France's Classic Grand Tours* (Somerset, UK: Haynes Publishing, 1990).

Stuck, Hans, *Männer hinter Motoren: Ein Rennfahrer Erzählt* (Berlin: Drei Masken Verlag, 1935).

Stuck, Hans, and E. G. Burggaller, *Motoring Sport* (London: G. T. Foulis & Co., 1937).

Symons, H. E., *Monte Carlo Rally* (London: Methuen & Co., 1936).

Taruffi, Piero, *The Technique of Motor Racing* (Cambridge, MA: Robert Bentley, 1961).

Taylor, Blain, *Hitler's Engineers: Fritz Todt and Albert Speer — Master Builders of the Third Reich* (Philadelphia: Casemate, 2010).

Tissot, Jean-Paul, *Figoni Delahaye: 1934–1954, La Haute Couture Automobile* (Anthony, France: ETA, 2013).

Tragatsch, Erwin, *Die Grossen Rennjahre 1919–1939* (Stuttgart: Hallwag Verlag, 1973).

Tucoo-Chala, Pierre, *Histoire de Pau* (Toulouse: Univers de la France, 2000).

Venables, David, *First Among Champions: The Alfa Romeo Grand Prix* (Somerset, UK: Haynes Publishing, 2000).

———. *French Racing Blue: Drivers, Cars, and Triumphs of French Motor Racing* (London: Ian Allan Publishing, 2009).

Walkerly, Rodney, *Grand Prix 1934–39* (Abingdon, UK: Motor Racing Publications, 1948).

Walter, Gerard, *Paris Under the Occupation* (New York: Orion Press, 1960).

Weber, Eugen, *The Hollow Years: France in the 1930s* (New York: W. W. Norton, 1994)

Weitz, Eric, *Weimar Germany* (Princeton, NJ: Princeton University Press, 2007).

Wolff, Marion, *The Shrinking Circle: Memories of Nazi Berlin, 1933–39* (New York: UAHC Press, 1989).

Yates, Brock, *Ferrari: The Man, the Cars, the Races, the Machine* (New York: Doubleday, 1991).

Zagari, Franco, *Tazio Nuvolari* (Milan: Automobilia, 1992).

Periodicals / Newspapers

ACF Magazine (French)

Alfa Corse Magazine (Italian)

Auto Age

Auto Moto Retro (French)

Auto Passion (French)

Auto Retro

Autocar

Automobile Quarterly

Automobilia (French)

Autosport

Bugantics

Car and Driver

Das Auto (German)

Delahaye Club Bulletin (French)

Englebert (French)

Horizons

Il Littoriale (Italian)

L'Actualité Automobile (French)

L'Auto (French)

L'Auto Italiana (Italian)

L'Automobile sur la Côte d'Azur (French)

L'Équipe (French)

L'Intransigent (French)

La Dépêche (French)

La Locomotion Automobile (French)

La Vie Automobile (French)

La Fanatique de L'Automobile (French)

Le Journal (French)

Light Car

Locomotive Engineer's Journal

Match (French)

Motor

Motor (German)

Motor Italia (Italian)

Motor Sport

Motor und Sport (German)

Motorspeed

Old Cars

Omnia (French)

Paris Soir (French)

Pur Sang

RACI (Italian)

Revue Automobile Club Feminin (French)

Road and Track

Speed

Sport Review's Motorspeed

Sports Car Graphic

Sports Car Guide

Sports Car Illustrated

Torque

Veteran and Vintage

Index

Page numbers in *italics* indicate illustrations and maps.

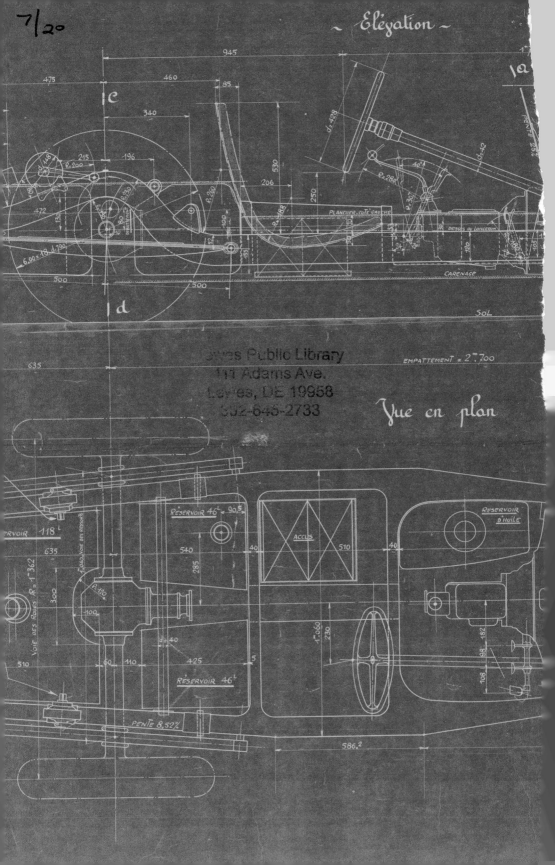

Vue en plan